Flying With 40 Horses

A History of the Continental A-40 Aircraft Engine and the Planes it Flew

By Chester L. Peek

Published By:
Three Peaks Publishing
Norman, OK

First Edition, First Printing
Printed in the United States of America
ISBN: 1-886196-04-4

Other books by this Author:

The Taylorcraft Story
The Spartan Story
Resurrection of a Jenny
The First Cub

Contents

Prologue

This book describes a search for power. Horsepower. When the early aviation pioneers such as Lilienthal, Chanute and the Wrights were carrying on their heavier-than-air flying experiments, they solved the mystery of flight using gliders. Guided by their wind tunnel data, the Wrights actually constructed a controllable glider and made over 1000 flights by the end of 1902. But they and their contemporaries knew that to make true flight possible, their machines needed a source of power. Thus the search for a lightweight engine, "a pound of feathers".

The Wright brothers had calculated, correctly, that they needed at least eight horsepower to drive their "flier" through the air. But at this time, the turn of the century, internal combustion engines were still in their infancy. Automobiles were just appearing on American roads, usually powered by heavy and cumbersome one or two cylinder motors. The Wrights, after much searching, concluded they would have to build their own engine. Their first primitive design, constructed by their expert machinist, Charlie Taylor, weighed about 200 pounds and produced twelve horsepower. It was just enough to lift their first successful "fliers" into the air.

By 1910, aviators in both America and Europe were matching the Wright's exploits, but all were plagued by the lack of suitable engines. Soon, World War I would bring on rapid technical development in this field, so that by 1918, four hundred horse-power motors were in common use. But, with the possible exception of the water-cooled Curtiss OX-5, no engines available were suitable for light civilian aircraft. Even after Lindbergh flew the Atlantic, and created the 1927-1930 "aviation boom", few small engines were being produced. Aeronca tried to meet the demand with their little two-cylinder model, but restricted its use to their own planes.

At this point in history, 1931, the Continental Motors Company introduced their famous Model A-40 flat four lightplane engine to the aviation industry. It met the special needs of those depression times. It was light, yet it was reliable, low in cost, and inexpensive to operate. A dozen lightplanes, including the famous Taylor Cub and the Taylorcraft, were powered by the A-40. Although it was ultimately replaced by its higher powered descendants in 1938, it left an enduring legacy. Its compact flat-four design became the standard for American aircraft engines, the engines that, even today, dominate the world market.

This book pays tribute to the Continental A-40, the gutsy little engine that put America's lightplanes into the air.

Chester L. Peek
Norman OK, 2001

Chapter One

The Search For a Pound of Feathers

The Wright Brother's engine. Very primitive, but it furnished the twelve horsepower needed to fly their first plane. HYDE

The Wright brothers needed an engine!

For three years this remarkably talented pair of brothers had been making successful flights in unpowered gliders. By careful experimentation, they had conquered the mystery of heavier-than-air flying, and had constructed a safe, controllable flying machine. But they needed power to make their invention practical. They had actually learned to fly by guiding their primitive gliders down the hills near Kitty Hawk, NC, recording over a thousand short flights in the fall of 1902 alone. However, in order for their machine to sustain flight and travel any distance, an engine was clearly needed.

Although America was in the midst of an industrial revolution, and the internal combustion engine was being widely used in the new autos, none were suitable for the powering of an aircraft. In December of 1902 the Wrights sent letters to a number of automobile companies and motor manufacturers, asking if an internal combustion engine could be furnished having at least eight brake horsepower and weighing less than 200 pounds. Significantly, one of the requirements was that it be "free from vibration" They realized that due to the fragile nature of their aircraft structures, severe vibration could be dangerous.

When no satisfactory replies were received, the brothers decided they would need to build a motor themselves. With their typical ingenuity, they produced an engine that weighed only 186 pounds and delivered between 12 and 16 horsepower. It was powerful enough to fly their first plane on that historic day, December 17, 1903.

The Wright's search for a suitable power plant illustrates the dilemma faced by American aircraft designers for the next thirty years. During World War 1, aircraft engine design progressed rapidly under the pressures of military competition. In 1914, a typical aircraft engine had perhaps 100 horsepower; by 1918, American planes were flying with the famous "Liberty" motor putting out a solid 400 HP.

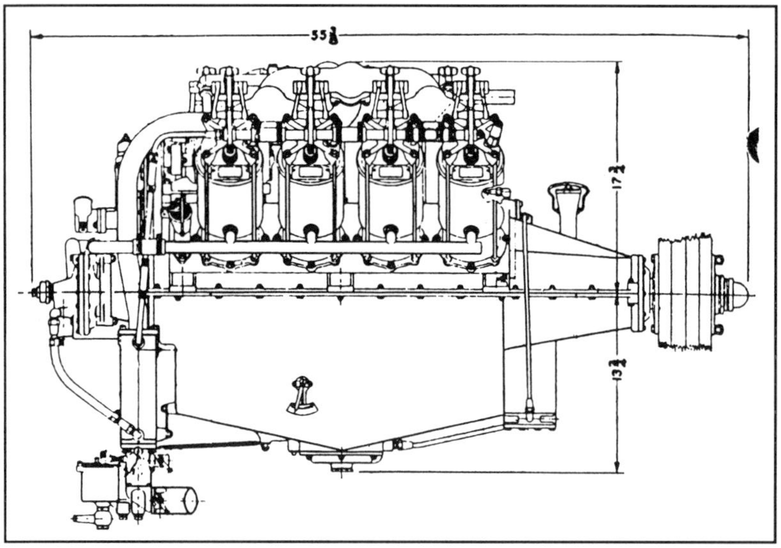

The venerable Curtiss OX-5 engine which powered most American planes in the 1920's CURTISS

A few tentative efforts had been made to meet this need. During World War I, Charles L. Lawrance founded the Lawrance Aero-Engine Corporation with the intention of developing a flat-twin air cooled motor. In one room on Broadway in New York City, he designed and produced the prototype of his two-cylinder 28 HP Model A. It weighed almost 200 pounds and, incredibly, was arranged so that both pistons worked off a single throw crankshaft! Lawrance evidently ignored the principle of physics that "each action causes an equal and opposite reaction". Thus, when both pistons in his engine went one way, the rest of the engine was forced in the other direction. To make matters worse, the power strokes were spaced unevenly, 540 degrees and 180 degrees.

Nevertheless, under wartime emergency conditions, 450 of these motors were produced under license by the Excelsior Motorcycle Company and installed in the Breese Penguin ground trainers. These clipped-wing machines were only used for taxi instruction, they were incapable of flight, so the engine problems were not considered dangerous.

After the end of the Great War, some of these engines appeared on the surplus market, selling for $75-$125. When Swen Swanson, (of Kari-Keen fame), designed his first single seat biplane in the University of South Dakota shops, he used one of these Lawrance engines. The plane actually flew! One enterprising mechanic, a Mr. Orville Hickman, devised a modification of the Lawrance A-2 engine which supposedly cured its destructive vibration problems. A two-throw crankshaft was substituted, and the connecting rods modified to compensate for the offset crank throws. Early would-be aircraft builders were desperate for any kind of lightweight power plant that would put them in the air.

But the development of higher horsepower engines for military planes did little to further the advancement of privately owned light aircraft. True, after WW I, the surplus Jennys and Standards offered acceptable mounts for the returning pilots who became the famous "barnstormers". Later, powered by the venerable and cheap but hopelessly obsolete Curtiss OX-5, the new production Wacos, American Eagles and Travel Airs met similar needs. Throughout most of the 1920's, no suitable small engine was available to power a really "light" airplane, one that could be bought and operated by the average flying enthusiast.

Swen Swanson's Lawrance-powered plane. U OF SOUTH DAKOTA

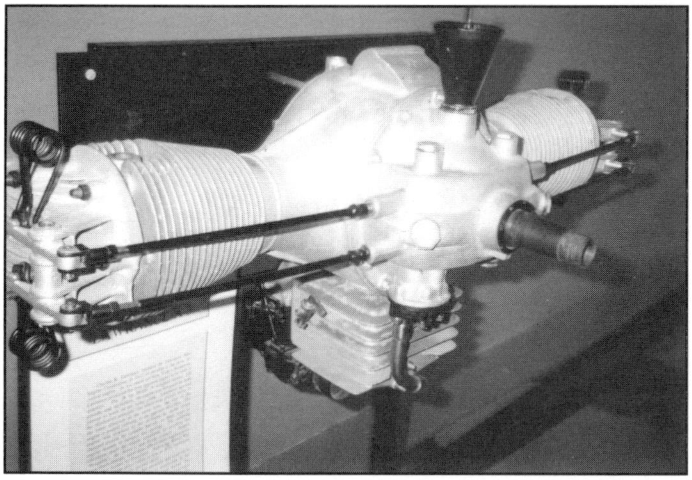

Lawrance's World War 1 trainer engine AUTHOR

Lawrance's next engine was a 3-cylinder radial used in the diminutive Sperry Messenger biplane, the L-1 of 223 cubic inch displacement and 56 HP. Next came a 9-cylinder radial, the R-1, 675 cubic inch , 150 HP.

In 1924, under pressure from high navy officials, a merger was arranged between Wright and Lawrance, Wright having the finances and production facilities needed for volume manufacturing. A new and larger nine-cylinder radial design was introduced, the Lawrance J-2, with 787 cubic inches of displacement putting out 200 HP. After some useful modifications by the brilliant British engineer, Samuel Heron, the Lawrance J-2 became the J-5 Wright Whirlwind. It achieved its initial, (and lasting,) fame when used by Charles Lindbergh to fly the Atlantic in 1927. But its cost was over $5000 and so it could in no way be considered suitable for a lightplane. Lawrance went on to become president of the Curtiss-Wright Aeronautical Corporation.

The Wright company did, however, make one attempt to satisfy this lightplane market. In 1924, Harold Morehouse was working as an engine designer and project engineer with the Army Air Service at McCook Field, near Dayton, Ohio. His specific project was to design a small air-cooled engine for use with army blimps. At the same time, two other McCook Field employees, Jean Roche and John Dohse were building a diminutive experimental monoplane in their garage. When they could find no suitable engine for their plane, they turned to Morehouse for help. An agreement was reached whereby the three would form a partnership to build a small two-cylinder motor of approximately 25-30 horsepower. Progress was slow, but by September of 1925 the first test flights were being made.

Both the little plane and the engine performed satisfactorily. Over the winter, over 200 flights were made, many by noted pilots of that area. Producing 29 HP at 2500 RPM, the motor was considered something of a breakthrough in the light aircraft field.

This caught the attention of Guy Vaughan, a Vice President of the Wright Aeronautical Corporation. In short order he hired Morehouse and purchased the rights to his engine design. (The original engine remained with the Roche and Dohse plane which later became the Aeronca. More on it later.)

With the resources of a large company at his disposal, Morehouse quickly refined his original design and by August of 1926, the Wright-Morehouse powerplant was being introduced to the trade. When announcing the new engine, Vaughan correctly assessed the current light aircraft situation in the United States:

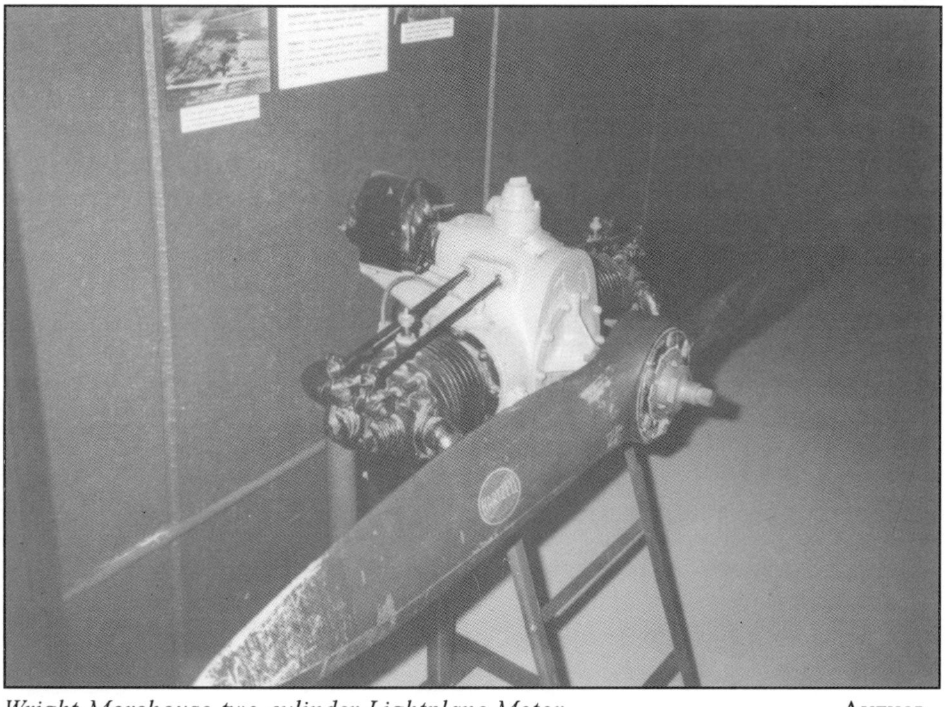

Wright-Morehouse two-cylinder Lightplane Motor AUTHOR

The original Roche monoplane, which became the Aeronca C-2. BALDWIN

"There is one phase of aeronautical development that has attracted a great deal of interest but has made very little real progress to date. It is the production of a really successful light plane. It has been claimed that the light plane is destined to become the flivver plane of the future; it will be the type that the average man will own and operate for pleasure and personal transportation.

"It is believed that the principal deterrent to this development is the lack of a suitable powerplant."

The Wright-Morehouse was an opposed two cylinder motor, quite modern in appearance. Somewhat heavier than its predecessor, it was still rated 29 HP at 2500 RPM. An examination of the photo on the preceding page indicates a sensible design, one that could probably be expanded into larger 4-cylinder models without difficulty. One clever feature; the lower half of the cylinder walls were cooled by an oil jacket. Although an initial production run of ten units was made, no sales were forthcoming, and Wright ceased its efforts to develop low horsepower engines. (We will next hear of Harold Morehouse when he joins Continental in the mid 1930's, and designs the famous A-50, A-65 series of engines).

While the Wright-Morehouse effort was not a success; the original engine, left with Roche and Dohse, would become the ancestor of the well-known Aeronca series of motors.

After Morehouse left the McCook Field trio, the two remaining partners continued to fly their little monoplane. But one day in early 1926 the airplane was involved in a serious accident, which injured Dohse and damaged both the engine and plane. Two more of their associates, Roy Poole and Robert Galloway, agreed to design and build another engine, based on the original Morehouse pattern. The model they produced was similar to the original, but somewhat larger and heavier. It performed well enabling the Roche monoplane to keep flying.

This Poole-Galloway engine would evolve into the Aeronca Company's Model E-107, the 107 indicating the cubic inch displacement.

Aeronca E-113 engine. BALDWIN

AERONCA E-113B

Aeronautical Corporation of America, Cincinnati, Ohio

• President: Taylor Stanley. Vice-President: H. V. Fetick, General Manager: C. S. McKenzie, Jr. Sales Manager: J. C. Welsch. Chief Engineer: Roger E. Schlemmer. Export representatives: Howard T. Byler, Rio de Janeiro, Brazil. Strachan & MacMurray, Iloilo, Philippine Islands.

Two-cylinder, horizontally opposed, direct-drive, air-cooled. Rated by the manufacturer at 40 h.p. at 2500 r.p.m.; by the Department of Commerce at 36 h.p. at 2400 r.p.m. ATC 71. Pressure feed lubrication to connecting rod bearings; splash feed to balance of engine; major bearings anti-friction. Compression ratio 5:4. Bore 4.25 in. Stroke 4 in. Displacement 113.5 cu. in. Length overall 24.125 in. Height overall 26 in. Width overall 34.5 in. Fuel consumption .55 lb./b.h.p./hr. Oil consumption .25 pt./hr. Octane fuel rating 73.

Construction: Crankshaft 5% nickel steel forging, counterbalanced. Crankcase heat treated "B" alloy. Ball bearings for thrust, rocker arm, camshaft; roller bearings on crankshaft; needle roller bearings on crankpin connecting rods. Piston Permite aluminum alloy, three rings to each cylinder, one oil, two compression. Cylinder heads heat treated Y alloy castings. Cylinder barrels heat treated steel forgings. Aluminum Industries exhaust and intake valves.

Accessories: Bosch magneto, Zenith carburetor, B. G. spark plugs, Cuno oil filter, screen air filter, automatic spark control.

Aeronca E-113B engine specs — *Aero Digest*

In the glow of the Lindbergh-induced aviation boom, investors all over the country were looking at aircraft manufacturing possibilities. The Aeronca company, of Cincinnati, backed by the Taft family fortunes, was one such organization. In May of 1929 they made a tentative offer to Roche. After a very enthusiastic test report by Major Gerald Brower, they finalized the agreement and Roche moved to Cincinnati. He received 220 shares of stock in the new company for the design and production rights of his plane and engine. After a short time, Galloway and Poole joined him in the new organization.

Even in those depression times the Aeronca became a modest success, 90 of the little planes were sold in 1930. The E-107 engine, producing 26-29 HP at 2,500 RPM, was reasonably reliable and inexpensive to operate. The engine parts were manufactured by the Grovo-Nelson company, but assembled in the Aeronca plant. In 1931, in order to have power enough to carry two people, the engine was slightly enlarged, and changed to an overhead valve configuration. It then became the 113-A, which, according to the Department of Commerce, produced 36 HP at 2400 RPM. The Grovo-Nelson people helped with the design, suggesting the use of Warner cylinders and heads, which they were already manufacturing. Later, twin magnetos were added and the horsepower increased to 40.

Thus the Aeronca 113 became one of the first true American lightplane engines manufactured in quantity. While reliable and inexpensive, it did not find a wide market, because Aeronca refused to sell it to other aircraft firms. At that time, (1930), small aircraft manufacturers had a very few suitable powerplant options available.

One such option was the Szekely motor made in Holland, Michigan. Otto E. Szekely, (pronounced ZAY-KI), was of Austrian descent, having graduated from technical schools in Vienna and Berlin. In the early 1920's he immigrated to the United States and joined the Velie Motor Car Co. of Moline, Illinois, as a mechanical designer. Later he formed his own consulting engineering firm, working for such companies as Maytag in Iowa and Cushman in Nebraska. It is interesting to note that all these companies were experimenting with small air-cooled engines at the time.

A Szekely engine intalled in a restored American Eaglet. Note the cables that hold the heads in place. AUTHOR

SZEKELY

SR-3-0

Szekely Aircraft & Engine Company
Holland, Michigan

Specifications: Length overall (less starter), 25 inches. Diameter overall, 36 inches. Mounting bolt diameter, 10-15/16 inches. Bore, 4⅛ inches. Stroke, 4¾ inches. Displacement, 190 cubic inches. Weight dry (less hub and starter), 150 pounds. Fuel consumption, .62 pounds per horsepower. Oil consumption, .009 pounds per horsepower.

The Szekely SR-3-0 aircraft engine is an overhead-type aircraft powerplant of 45 horsepower at 1,750 revolutions per minute. It is an air-cooled radial-type three-cylinder four-cycle engine. Compression ratio is 4.6. Dual magnetos, either Bosch or Scintilla are provided. Carburetor is Stromberg. The lubrication system involves gear pumps, pressure and scavenger. There is one intake and one exhaust valve per cylinder. The propeller drive is direct.

The pistons are aluminum alloy and there are three rings per piston. The cylinder heads are aluminum alloy. The cylinder barrels are of cast iron molybdenum. The crankcase is aluminum alloy.

The Szekely SR-3-0 is a slow speed engine, operating most efficiently at between 1,500 and 1,800 revolutions per minute, with an average cruising speed of approximately 1,600 revolutions per minute, according to officials of the Szekely company, on the principle that slow speed will result in longer life and better service. This engine is designed particularly for light low-priced aircraft.

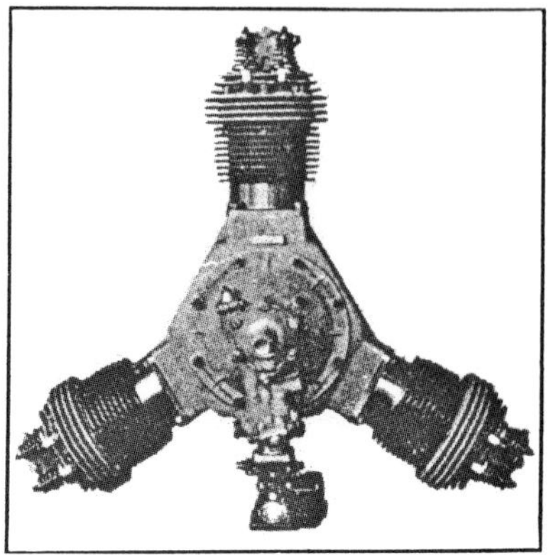

Szekely Specifications AERO DIGEST

In 1925, at the urging of the Holland Furnace Company, Szekely moved his operations to Holland, Michigan, where he was involved with such projects as washing machine and piston ring manufacturing. In 1928, he became interested in the aviation market and designed a small, three-cylinder radial aircraft engine, the L-head SR-3-L of 30 HP. Szekely even tried his hand at aircraft manufacturing, producing several single seat mid-wing monoplanes, named, appropriately for its location, "The Flying Dutchman".

At least 10 lightplanes received their ATC's using the Szekely engines including such well known names as the Curtiss-Wright Junior, Buhl Pup, Rearwin and American Eaglet. In fact, the American Eaglet became the first plane in the United states licensed to carry two people, (using the Szekely SR-3-L), with only 30 HP. The little engine cost less than $600, and for a time was one of the few options available to the light aviation trade. While the Szekely had a number of mechanical problems, including moderate vibration, it did deliver the horsepower needed with a reasonably low weight.

Although they were installed in a number of lightplanes, and perhaps a thousand engines were built, the enterprise was not a success. The company filed for bankruptcy on May 10, 1932. For several years thereafter, engines could be purchased at "close out" prices from the company, and were still being installed in new planes as late as 1935.

The Light Manufacturing and Foundry Co. of Pottstown, Pennsylvania, made a valiant, if futile, effort to enter the small engine field in 1929. They had been introduced to this market when they began to do machine and assembly work for Brownback Motor Laboratories of New York City. H. L. Brownback, the owner, had been acting as the American agent for the French Anzani company for several years. In 1928 he induced the Light Foundry to build complete engines, based on an almost exact copy of the Anzani design.

A Brownback Tiger 6-cylinder radial powers this barnstorming Kari-Keen Coupe near Paullina, IA, 1940 BOCK

This low-priced American Eaglet entry into the lightplane field hoped to use the very small Cleone 2-cycle 25 HP motor to reduce cost. Unfortunately, even though the price was low, the performance was not satisfactory, so it was withdrawn from the market. Reengined with the Szekely SR-3-L, it was a moderate success. — Aero Digest

LIGHT

TIGER KITTEN 20 AND 30

Light Manufacturing and Foundry Company
Pottstown, Pennsylvania

Specifications: (Models 20 and 30 respectively); 24½ inches; 28 inches. Height overall, 17¼ inches. Width overall, 12 inches. Bore, 3¾ inches. Stroke, 3½ inches. Displacement, 77 cubic inches. Weights dry, 80; 95 pounds. Fuel and oil consumption, .65 pounds per horsepower per hour.

The Light Tiger Kitten 20 is a two-cylinder air-cooled in-line inverted aircraft powerplant developing 20 brake horsepower at 2,400 revolutions per minute. The Tiger Kitten 30 is a two-cylinder geared air-cooled in-line inverted aircraft powerplant developing 25 brake horsepower at 3,000 crankshaft revolutions per minute; 1,500 propeller revolutions per minute. Compression ratio on both models is 5.3. Both models have a single carburetor and a single magneto. The lubrication system is forced feed throughout. The cylinder heads are Y alloy, cylinder barrels cast-nickel iron, with integral fins, bolted to crankcase with long hold down bolts passing through head. The pistons are Y alloy and there are five rings per piston. Connecting rods are forged steel "I" section, roller bearing. The crankshaft is two-throw, one piece. The crankcase is aluminum alloy No. 195, heat treated, split on center line of shaft.

There are two power impulses per propeller revolution on Model 20 and four power impulses per revolution on Model 30. There are eight moving parts on Model 20, and 10 moving parts on Model 30.

Specifications for the Light "Tiger Kitten" engine AERO DIGEST

One of their first products was the "Brownback Tiger", a two-row air cooled radial rated at 90 HP. It was smooth running, but quite heavy for the power delivered, (278 pounds). The Department of Commerce issued ATC #33 for this engine. A small production run was made and several aircraft manufacturers, (including C. G. Taylor who was then making the ill-fated Chummy), tried the engine. The Kari-Keen Sioux Coupe, made in Sioux City, Iowa, apparently was the only plane to receive an ATC with this motor, #413, dated 4-8-31.

When their medium-horsepower radial engines failed to sell, the Light Foundry people elected to introduce even smaller models, the Tiger Kitten 20 and 30, intended to power the very lightest planes then being produced. Their design departed from the conventional practices by using the two-cycle principle, and in the case of the Tiger 30, the engine was geared down two to one.

It appears the Light Company had an excellent opportunity to become a major player in the aircraft engine field, but missed the chance. They were the very first motor tried in C. G. Taylor's prototype "Cub", the plane that was destined to revolutionize the light aircraft field. C. G. Taylor remembered the first trial of the Tiger Kitten:

"A man named Kirkendal flew up from Pottstown with a little two-cycle engine in the front seat of a Fleet biplane. I knew it wouldn't fly the plane, but he was persistent. Since nothing else was available, we made a temporary engine mount and let Kirkendal try to fly it. The plane would just run down the field, only getting off the ground when it hit a bump. It was underpowered and just wouldn't fly the plane."

The Tiger Kitten engine achieved some measure of immortality, however, when it became the inspiration for the famous "Cub" name. Again, according to C. G. Taylor: *"As I recall, we still did not have a name for the little plane, and we couldn't call it a Chummy because it was a tandem seating arrangement. So when Ted Wells suggested 'Taylor Cub sounds good with Tiger Kitten', we used it".*

Auto engine conversions would offer another possible source of inexpensive horsepower for the light aircraft builder. One of the earliest attempts at such an application may have been made by one Charles Van Auken. In 1909 he is said to have installed one of the first Ford Model "T" engines in a flier of his own design. Unfortunately, soon after the contraption left the ground the crankshaft broke and a crash ensued.

Years later, a quite successful auto engine conversion was marketed by Bernie Pietenpol, of Cherry Grove, in southeast Minnesota. A farmer of Dutch decent, Pietenpol was smitten with the flying bug in the 1920's. Having no funds with which to buy a plane, he decided to build his own. Showing a good bit of native ingenuity, he came up with a high wing monoplane design that carried two people. For the most part it was of simple wood construction, making it an ideal project for the inexperienced homebuilder of the 1920-30 era.

Bernie Pietenpol's Air Camper AUTHOR

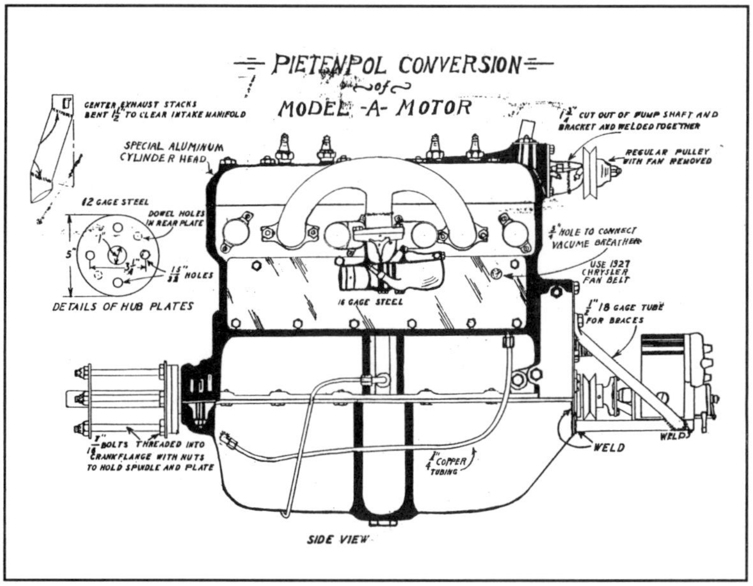

Pietenpol's Ford Model "A" engine conversion PIETENPOL

Best of all, it was powered by a converted Model "A" Ford motor. These products of Henry Ford's genius were available throughout the country at any salvage yard. Pietenpol even published plans showing how to build the plane and convert the engine for aircraft use. Over five thousand sets of these plans were sold, and a fair number of planes were completed.

Bernie Peitenpol had no formal engineering training, so his approach to the design was intuitive, not scientific. One story mentions his first attempt at wing design. He had used an airfoil that was scaled down from the Curtiss Jenny, very thin and flat. According to Bernie, when he put the wing on the plane, *"It would just run up and down my pasture, but would not lift off the ground."* So he built a new wing with a *"fattened up"* airfoil and the plane flew fine. Some reported it would haul two people on a hot day better than a 40 HP Cub.

The Model A engine chosen by Peitenpol was, of course, quite heavy. But it was cheap and it was reliable. It was easily repaired by any auto mechanic, and spare parts were available everywhere. During the stressed economic times of the Great Depression, it offered the fledgling flier an economical way to experience flight.

Surprisingly, a number of these planes are still being built and flown today. Most adhere closely to Bernie's original design, except in the choice of engine. While purists insist on the venerable Model A, many opt for the more modern and lighter Continental A-65. A few have even been built with the Corvair six-cylinder air-cooled engine. There is still an active Pietenpol club, with newsletters and photos available on the internet.

Several other lightplanes actually received Approved Type Certificates using conversions of this same Model A motor. One such plane was the Wiley Post, a small biplane built in Oklahoma City. This firm got its start when Frank Staughan, a wealthy auto dealer from Marshall, Oklahoma, began financing an aircraft designed by Ross Holmes. In 1932, Holmes set up a shop in a corner of the Cessna plant in Wichita, and built a small biplane. Glenn Stearman is reported to have done most of the engineering design work.

A line-up of Wiley Post biplanes at the Oklahoma City factory. JUPTNER

Marion McClure's Wiley Post biplane AUTHOR

The Model A powered prototype was completed in 1933 and was well liked by those who flew it. Plans were made to manufacture a number of planes in the former Yellow Aircab plant, but financial difficulties soon arose, shutting down production. The company was bought the next year by Oklahoma City oil man Mark Kleedon, and the entire operation moved to the Wiley Post Airport in that city. Burrell Tibbs was hired to set up production, and, to capitalize on his famous name, Wiley Post was named president. But the planned sales volume did not materialize, and the plant was shut down after making only thirteen of the cute little biplanes.

The Author had the privilege of flying Marion McClure's Model A powered Wiley Post biplane back in the late 1950's. It was quite an experience. The cockpit of the little plane was arranged side-by-side, but with the right hand passenger seat about 6 inches behind the pilot to allow shoulder room in the narrow fuselage. The engine was unusually quiet and smooth, when compared with the typical 65 HP aircraft motor. On takeoff, the antique tachometer showed about 1800 RPM, but with the heavy engine block completely dampening all vibration, you had the impression it was not turning up. The plane got off the ground well, and although it seemed somewhat nose heavy, flew just fine. One could understand why the homebuilt Pietenpols with Model A power became so popular.

Another Ford-powered plane was built by the Akron Aircraft Company, of Akron, Ohio. Both the plane and engine were designed by two Kansas brothers, Joe and Howard Funk. They received ATC #715 on their little monoplane which had a much modified, and inverted, Ford Model B engine. It first appeared at the Miami air races in 1937, where it was praised by the famous trans-Atlantic flier, Clarence Chamberlain. Soon after this introduction, the Funk brothers received a development grant, ($7500), from the Director of Air Commerce, Eugene Vidal. One plane was to be furnished so officials could evaluate the performance of an auto engine for aircraft use. This encouraged several Akron investors to form the Akron Aircraft Co. and production was started in 1938.

The Funk "E" engine was rated at 63 HP at 2125 RPM, with 70 HP available at 2450 RPM on takeoff. The performance was generally satisfactory, although early units had minor cooling and lubrication problems. However, the flying public did not welcome this innovative powerplant, so by 1941, the Funk plane was made available with the Lycoming GO-145-D2 geared engine.

Perhaps the most inexpensive Ford-powered flying machine ever offered was produced by the Universal Aircraft Company of Fort Worth, Texas, and shown on the following page. Dubbed "The American Flea", presumably after the French Flying Flea popular in Europe, it was a diminutive triplane of unusual design. Only 13 ft. long and with a 20 ft. wingspan, it was certainly a bare minimum of airplane.

JUPTNER

An early Model B Funk. Note the hand crank on the side of the cowling

An American Flea triplane, powered by a Ford engine AUTHOR

The powerplant is somewhat of a mystery. It was advertised as a 40 HP air cooled motor, the Universal 40. Photos taken at the time indicate it was probably a Model A block, converted in some way to air or oil cooling. With a list price of only $85.00, it certainly could not have been too sophisticated a design.

The plane itself was equally cheap. The unfinished kit cost $99.00; covered and doped the price rose to $125.00. There is no way of determining how many of these motors and planes were produced, since they were undoubtedly not registered with the Department of Commerce. According to old-time flyers in the Dallas-Fort Worth area, Wayne Brewer did buy one of the little planes, but installed a Szekely engine.

In a desperate search for cheap horsepower, one enterprising young engineering student at the University of Washington, Gernet Heinemann, even converted a marine engine for flight. By the late 1920's, the Johnson Sea-Horse Company was offering a 32 HP outboard, some were being installed to power the midget half-mile dirt-track racers around the country.

This was enough to tempt a light aircraft builder. Even though the engines were two-cycle, and turned much too fast for use with a propeller, Heinemann designed a conversion and installed it in a Heath Parasol airframe. It was geared down 1:2 so that a large diameter propeller could be used. A special landing gear was designed to allow for proper ground clearance on take-off. A rather neatly cowled nose-type radiator was utilized for water cooling.

With the plane weighing only 300 pounds empty, we can assume the motor offered adequate performance. Reports are that the plane made frequent flights in the Seattle area.

Throughout the 1920's, hopeful amateur aircraft builders were also attempting to fly their small homebuilt creations with motorcycle engines. It seemed a natural choice. The motors were relatively light in weight and offered the simplicity of air cooling. But when installed in the smallest of airframes, they usually proved to have inadequate power and were often rough running. This was especially true of the most common types available, the Harley-Davidson and Indian V-Twins. One example is shown on the page following. Four-cylinder motorcycle engines, however, did offer more promise. They were smooth running, and when properly tuned, would put out up to 25 HP, theoretically enough to fly a small single seat plane.

JUMPTNER

This small biplane with a 21 foot wingspan and a weight of 315# was designed in 1923 by O. H. Snyder, an aeronautical engineer with the U. S. Army. It was powered by an Indian twin "Powerplus" motor and drove the propeller through a 2:1 chain drive reduction gear. AUTHOR

The Henderson Four was one of the premier motorcycles of that era. William G. Henderson, along with his parents and younger brother Thomas, immigrated from Scotland about 1880. Trained in the fledgling auto industry, he took an early interest in small air-cooled motors. In 1912 he designed a four-cylinder motorcycle which gained the attention of Detroit businessman, George Monangham. Sensing an investment opportunity, Monangham raised $175,000, leased a plant at 268 Jefferson Avenue, and started the Hendersons in the motorcycle business. The first years production of 1000 units sold out quickly, and by 1916 the "Henderson Four" was one of the most desired, best performing cycles in the country. Even though the price of $325 almost equalled that of a small car, cycle enthusiasts, especially those interested in racing, continued to purchase the machines.

Their success caught the attention of the Schwinn organization whose Excelsior motorcycles were almost as popular as the industry leaders, Indian and Harley-Davidson. Supposedly, a six-figure sum changed hands, and Henderson was joined with Schwinn. For a time things went well, but after about a year the Hendersons left, due, it was said, to design disagreements.

In 1919, the Henderson brothers again found financial backing and started a new company in Philadelphia, manufacturing a similar four-cylinder motorcycle named the "Ace". Over the next several years, both the Henderson and Ace companies continued to improve and refine their little four-cylinder in-line engines, until they produced an advertised 25 HP. At this point, about 1926, these engines began to attract the attention of small aircraft builders.

When Ed Heath designed his first single seat "Parasol", he installed a Henderson engine. It worked quite well. After all, his little monoplane was lightly built and Heath himself weighed only 110 pounds.

The 1931 Henderson Motorcycle, the last year of production. This engine had 80 cu. in. with overhead inlet valves and a five bearing crankshaft. AUTHOR

HEATH B-4 **4-cylinder—in-line** **APPROVAL E4-3**

Specifications. Direct drive, air-cooled. Department of Commerce rating 25 horsepower at 2800 revolutions per minute; manufacturer's rating 27 horsepower at 2700 revolutions per minute. Cruising rating 25 horsepower at 2450 revolutions per minute at 3000 feet; horsepower available for take-off 30, at 3000 revolutions per minute. Compression ratio 4.6:1. Bore, 2.75 inches; stroke 3.5 inches; displacement 83 cubic inches. Length 25.25 inches; height 25 inches; width 14 inches. Weight 109 pounds. Power weight [illegible] pounds per horsepower. Fuel consump- [illegible] .52 pounds per brake horsepower per hour. Oil consumption .006 pound per brake horsepower per hour. Fuel octane rating 70. **Construction.** Crankcase aluminum alloy; two piece. Crankshaft heat treated nickel steel; one piece; three bronze backed babbitt bearings; 1 Federal No. 305 ball thrust bearing. Connecting rods heat treated alloy steel. Cylinders L-head with cylinder head cast integral with bronze port plugs over valves. Cylinder barrels nickel iron castings. Pistons heat treated aluminum alloy with three piston rings for each cylinder. Pump double gear type lubrication system with pressure distribution to all bearings. Tachometer drive from rear of camshaft. Camshaft heat treated aluminum alloy on three plain bearings and one ball thrust bearing. **Accessories.** Bosch magneto, Zenith carburetor, Heath starter, B. G. spark plugs, Thompson valves.

Heath-Henderson motor specs AERO DIGEST

In 1927, Ed Heath began to sell plans and kits for his Parasol, and a demand for engines developed. Heath answered this need by developing the Heath-Henderson motorcycle engine conversion. A number of changes were made, all in an effort to give the engine more power and reliability. A complete converted engine was offered for $250.00.

Alternately, conversion parts could be bought from the Heath factory if the customer wanted to modify his own engine. These parts included:

Nose casting, shaft & bearing....$10
Prop hub.......................................$8
Heath propeller...........................$18
Deep crankcase...........................$25
Set, alum port plugs.....................$6

The engine weighed 115 pounds. It was supposed to develop 27 HP at 3000 RPM, but that figure is open to some question.

As more of the engines came into use, the Heath people found that a stock motorcycle engine would not stand up under the nearly full power operations demanded for flight. With only 5-10 lbs oil pressure, the lower end bearings soon burned out. This was remedied by installing a high capacity oil pump and using tight fitting, closed type bearings.

Next, the exhaust valves on the back cylinders started to burn out after only a few hours use. Extended aluminum port plugs were designed which took the excess heat away from the valves. The overheating of the third cylinder was handled in a novel manner. A lower compression piston was installed on this one cylinder, reducing its heat output.

Flight experience soon proved that the regular Henderson lower crankcase was not suitable for aircraft use. It had been made shallow and heavy on the bottom, due to its proximity to the road when installed in a motorcycle. It only held two quarts of oil. Thus, when the plane nosed over in a glide, oil would flow forward and come out the breather, covering the cowling with oil. Conversely, when in a steep climb, there would be a loss of oil pressure because the suction line at the front of the case would be starved of oil. This would often cause bearing failures.

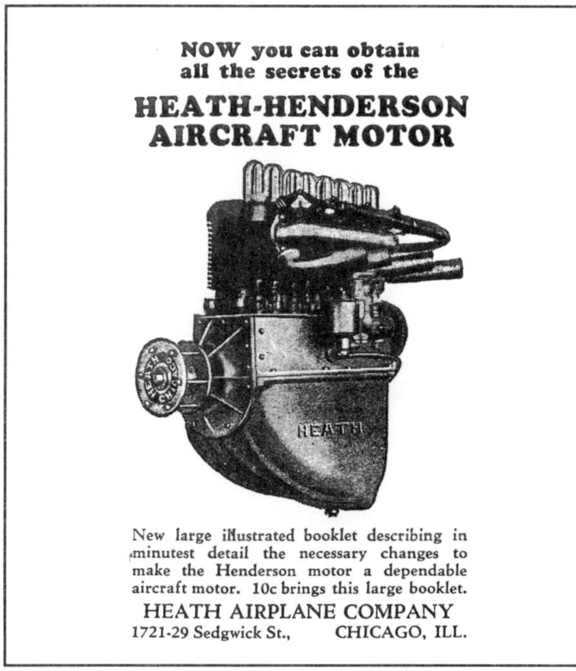

Heath-Henderson motor ad AUTHOR

Heath cured this problem by designing a special deep crankcase holding six quarts of oil. This assured a plentiful supply of lubricant to the bearings in all flight attitudes, and aided in the cooling as well.

After this developmental period, the Heath company was justifiably proud of their little engine. Their advertising brochure contained this statement:

"After many years of experimenting and constant use, both on the ground and in the air, we finally have brought this four-cylinder motor to such a high state of perfection that it will rank with any of the modern aircraft motors for reliability and economy. Thus we have perfected the most practical lightplane motor ever offered at a price within the reach of all. The weight per horsepower output is comparable with many modern motors, and the fact that the motor offers little head resistance gives an added performance. We have found that in planes of like design the top speed is equal to that of a motor having at least 10% higher horsepower due to the lower head resistance offered by the four-in-line motor".

This midget biplane was powered by a Bristol "Cherub" in the 1926 Philadelphia National air races. JUPTNER

The real truth was that the Heath engine, like its competitors, was not really satisfactory for sustained, day after day, flying. Added to that, their supply of used motorcycle engines was drying up. Very few Henderson cycles were made after 1929, and production ceased forever in 1931.

European powerplants were also available to the lightplane builder, but none appeared to be attractive options. We will read later, in Chapter 4, that C. G. Taylor used an imported French Salmson radial engine in his first Cub. It performed well, but was expensive and supplies were limited.

In 1923, the Bristol Company in England introduced their diminutive two cylinder opposed "Cherub". It put out 25 HP at 2500 RPM, weighed only 90 pounds, and had 67 cubic inches of displacement. Later a more powerful geared version was offered which produced 36 HP at 3200 RPM. It was a complex, well built little powerplant, with overhead valves and steel liners in aluminum cylinders. But again it's $1290 price tag kept it from becoming a major factor in the industry. The midget racer shown above, used this engine.

****

From the Wright brothers first flight in 1903 until the aviation "boom" year of 1929, American aircraft builders had struggled to put their planes in the air using an array of inadequate, unreliable and expensive aero engines.

But this deplorable situation would soon be remedied. An industrial giant, Continental Motors, although wounded badly by the depression, was about to enter this neglected field. When they did, the history of light aircraft in the United States would be changed forever!

Chapter Two

Continental's A-40

This is an artist's rendering of the original Continental A-40 as it appeared in their first color brochure, February, 1931. Note the heads do not have the deep cooling fins of the later models, have only 11 hold down studs, and there are no finned aluminum exhaust fittings. WHITTIER

By the Spring of 1930, America's light aircraft engine industry, if you could call it an industry, was in a shambles. The only powerplants available to the lightplane manufacturer were a mishmash of unreliable and expensive radials, or equally unattractive converted auto and motorcycle engines. The ancient, if cheap, Curtiss OX-5 war surplus motors were almost gone. The lack of suitable power was stifling the few remaining light aircraft builders.

Then, when the outlook was the darkest, Continental's A-40 appeared on the scene. Born of a desperate necessity during the Great Depression, it pioneered the efficient "flat opposed" design which still powers most of the lightplanes in the world today.

Continental Motors Corporation had a long and illustrious history. It was the brainchild of Ross W. Judson, a Minnesota native, who showed remarkable mechanical talents at an early age. When he graduated from the Armour Institute of Technology in Chicago in 1902, he was encouraged by his instructors to build a career around his interest in engines. Starting a machine shop in Chicago the same year, he introduced his first production motor to the trade in June. It was quite advanced for the day, a four-cylinder, four cycle model with an L-head arrangement.

His early products were well received, so in January of 1903, he formed a partnership with his brother-in-law, Arthur Tobin, who came from a wealthy Chicago family. Their business, supplying motors to America's fledgling automobile industry, grew rapidly. By 1904 they were advertising in national publications and had named their product "Continental Motors", indicating the European influences on their designs. Production that year reached 100 units.

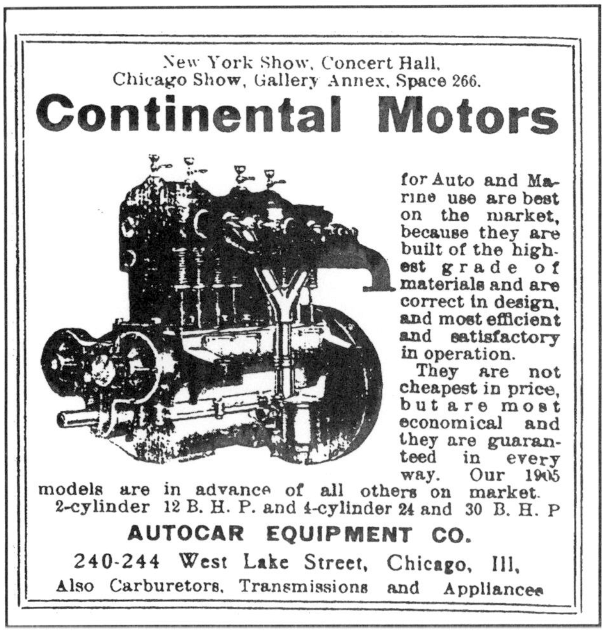

Early Continental ad, 1904 THE HORSELESS AGE

Business continued to grow in the early months of 1905. Continental Executives had already decided to concentrate their efforts on "motors, not cars". Since most of the early car makers did not make their own engines, this opened a vast market for Continental. In July of 1905, Judson and Tobin entered into a contract with the Muskegon Chamber of Commerce to move their operations there. They were offered a generous incentive arrangement: If they employed 125 men for six years, they would collect a $12,500 bonus, a large sum in those days. Municipalities used all sorts of inducements to attract manufacturing plants to their cities; it is still a common practice to this day.

The financial incentive was not the only reason for the move. Lower Michigan, specifically Detroit, was becoming the center of the auto industry. The move to Muskegon would put Continental much closer to its auto-making customers. That same year a two-story 16,000 square foot brick factory and office building was built, production starting there in May of 1906. Both their horsepower and sophistication of design made giant steps forward, but almost all motors were still refinements of their original four-cylinder model.

Sales continued to soar for the next few years, exceeding 15,000 units in 1910. In early 1911, the Hudson Motor Car Company placed an order for 10,000 engines, the largest single order ever received up to that time by an engine maker. This offered Continental another opportunity for expansion, and a move to Detroit, now the "motor city". Hudson sold them a site across the street from its auto assembly plant, and a long term relationship was cemented. The Muskegon plant had now grown to 110,000 square feet, but the new plant built in Detroit covered 225,000 square feet, tripling their factory space.

In his position as President and Chief Engineer, Judson guided the company in its first efforts toward automation. As early as 1908 he had installed a conveyor along which various heavy engine parts were moved on rollers from station to station for assembly operations. In an era when most motors were assembled with hand-fitted parts, he also pioneered the precision machining necessary for the standardization and interchangeability of all components.

To demonstrate this feature to the press, a unique experiment was devised. After 10 engines had been put through their initial test runs, engine workers were ordered to dismantle them and scramble the parts. The 10 engines were then reassembled, and all ran perfectly! They boasted it was a "first" for their industry.

Because of an increasing demand for higher horsepower and lighter weight, the company introduced its first six-cylinder model in 1912. Developing 45-60 HP, the cylinders were cast in threes, bringing them

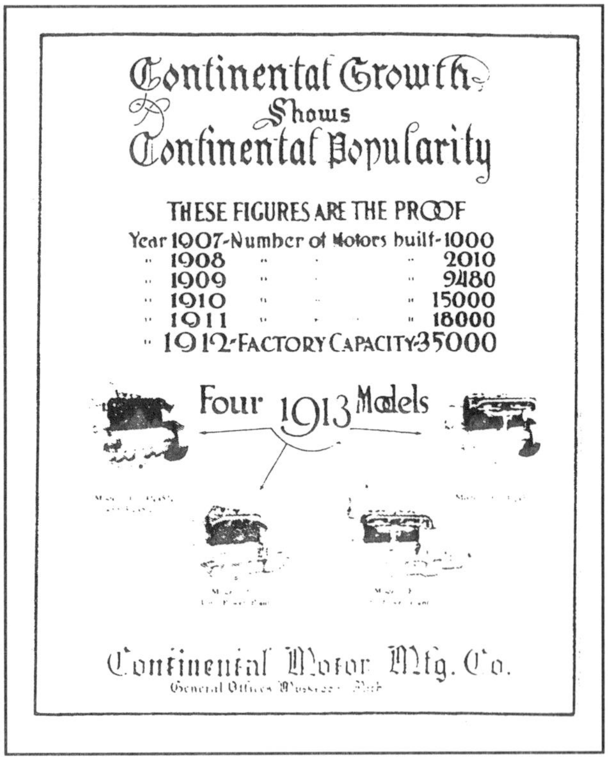

Continental advertises their growth AUTOMOBILE

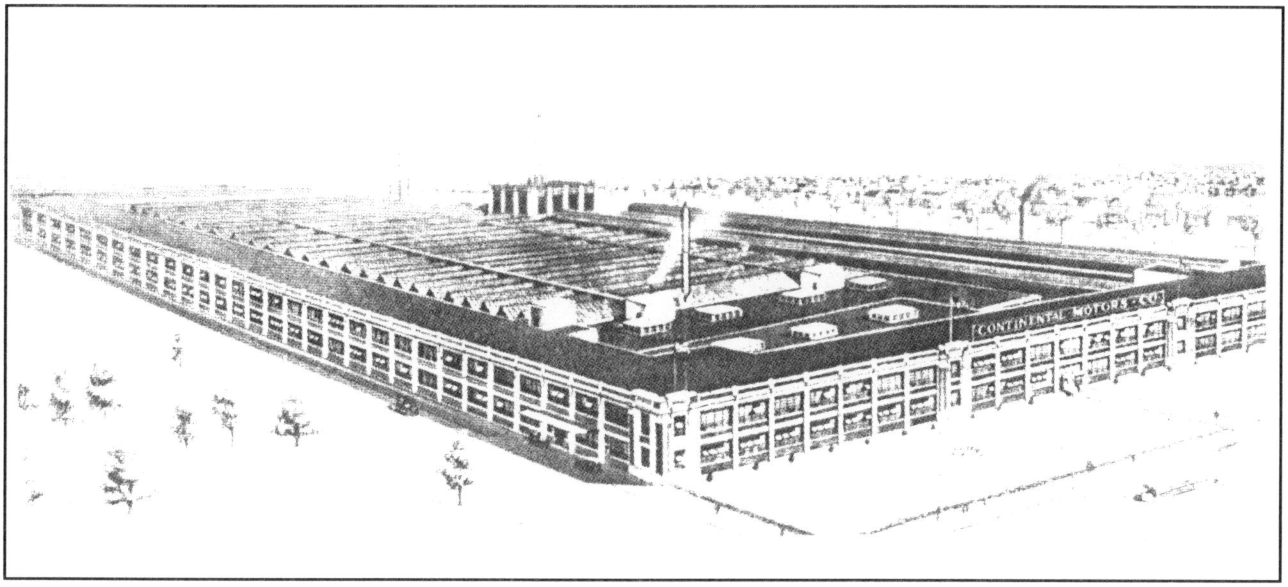

An artists rendering of Continental's huge Muskegon plant as seen in 1924 CONTINENTAL

closer together and reducing the length. The inherent smoothness of the six-cylinder arrangement pleased customers; more and more of their sales were of these models.

Continental's customer list for 1912-18, reads like an antique auto buff's dream. In addition to Hudson, their largest customer, they also sold motors to Auburn, Davis, Saxon, Star, Willys-Overland, Lexington, Abbot-Detroit, Velie, Moon, Elcar, Kissell, Columbia, Paige, Case, Comet and a host of others.

In 1917, the United State's entry into World War I ignited a boom in all forms of manufacturing, and especially anything associated with the auto industry. Continental did its part by consulting with the government on engine design, then turning over its Muskegon plant entirely to military truck engine production. Commercial and military output rose to a record high of 12,000 units per month.

Throughout the "roaring twenties", Continental continued to prosper, but there were danger signs on the horizon. The number of "independent" car makers was dwindling, the giants remaining tended to make their own engines. Some at Continental recognized this and began searching for alternate products or markets. A somewhat successful program was launched to build industrial engines, and forays were made into the military and agricultural areas. An alliance with Grey Marine opened the boating market to Continental.

But few of these initiatives did well. Back in 1917, William C. Angell, a Muskegon lawyer joined the corporation as Corporate secretary. As the years passed he took on more responsibilities, becoming Executive Vice-President in the mid-1920's. He was particularly interested in adding new or innovative products to Continental's product mix. On a trip to Europe in 1925, Angell became enamored with a Scottish designed sleeve valve engine. Hoping it would revolutionize the auto engine industry, he purchased worldwide patent rights and prepared to set up production in their Detroit plant.

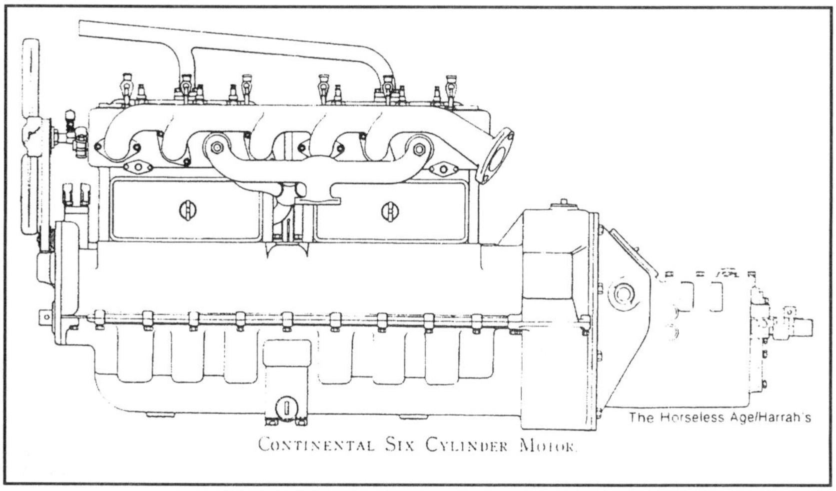

Typical six-cylinder motor, 1916 HORSELESS AGE

An early Continental A-70 ad AERO DIGEST

The advantage of the sleeve-valve design was said to be a simplified operation which tended to be quieter, and of greater durability. Other claims included greater economy, decreased weight, and a higher speed with less vibration. Continental went about making its first basic design change since the company was founded. At the request of Tony Fokker, they would also introduce a line of sleeve-valve aircraft engines.

In spite of Angell's optimism, and massive marketing efforts, the sleeve-valve venture was a complete failure. The auto manufacturers did not embrace this radical change in design, and the project was dropped. However, it did have one unintended consequence. Continental had decided to enter the aircraft engine field.

The company had been active in aviation for a number of years. As early as 1923, Judson had acquired a flying boat, and in 1926 the company purchased a Fokker tri-motored transport; using it for both nationwide executive travel and sales promotion activities. Their first attempt at aircraft engine design was actually a copy of the Wright motors used on their Fokker, with sleeve valve cylinders attached. Although shown at both the 1927 New York and Chicago Auto shows, this project never went beyond the prototype stage.

But when Lindbergh made his historic transatlantic flight in his "Spirit of St. Louis", interest in aviation, and the prospects of aviation engine manufacture, suddenly exploded throughout the country. In September of 1928, Engell announced the formation of Continental's Aircraft Engine division. To head this new venture, Robert Insley, an engineering graduate of Massachusetts Institute of Technology, was brought in. He had headed the Army Air Services Power Plant Branch at Dayton, Ohio for seven years. Not only was Insley an experienced engine designer, he also had the connections within the armed services that could lead to military sales.

The first design of the new division appeared the following spring; a seven cylinder air cooled radial, the Continental A-70, which produced 165 HP and would later become the famous W-670. It was first shown at the 1929 Cleveland air races. In October of that year the engine received its approved Type Certificate from the Department of Commerce, by which time over a dozen aircraft manufacturers had placed orders. But the expected large volume of repeat sales was not forthcoming. In fact, factory records show that less than 100 were sold from 1929 to 1931.

The stock market crash of 1929 had signaled the end of America's industrial "boom" years. By the Spring of 1930 Continental Motors Corporation was in serious trouble. Their 1929 fiscal year had ended with

Name of Engine:	**CONTINENTAL A-70.**
Manufactured by:	Continental Aircraft Engine Co., Detroit, Mich.
Type:	7 cylinder, radial, air cooled.
Rating:	(Dept. of Commerce Approved Certificate), 165 H.P. at 2000 R.P.M.
Displacement:	544 cu. in.
Compression Ratio:	5.4 to 1.
Dimensions:	Length, overall .... 39" Diameter, overall .... 41¼" Bore .... 4⅝" Stroke .... 4⅝"
Weight:	430 lbs. without hub or starter.
Fuel Consumption at Rated H.P.:	Not more than .55 lbs. per H.P. hr.
Oil Consumption:	Not more than .035 lbs. per H.P. hr.
Lubrication:	Pressure and scavenger gear pumps, 75 lbs. to the sq. in.
Ignition:	2 Scintilla Magnetos.
Carburetion:	Stromberg.
Spark Plugs:	2 per cyl. B. G.
Price:	$2,700.

A-70 specifications, price is $2700! AUTHOR

reduced profits, and the elimination of the customary dividend payment on common stock. In February of 1930, Ross Judson, the company's founder, had been replaced by Angell, who represented a large group of dissatisfied stockholders. Several other key executives, including Engineering Vice President Walter Frederick, left about this time. As the months went on, Continental's share of the auto engine market shrank to almost nothing. They had sold 330,000 engines in 1929; by 1931 they would make only 13,000.

The Continental management could now see that the market for the very expensive A-70 would be limited, even if government contracts were secured. Its $2700 price tag was just too high for those stark depression times. An aircraft engine that sold for hundreds, not thousands of dollars, was needed.

It was in this atmosphere that the famous Continental A-40 was conceived in early 1930. It would prove to be a pivotal event in the development of America's light aircraft industry.

Surprisingly, the design that emerged from Continental's engineering laboratories in no way resembled their considerably larger A-70. This was due to several factors. First the new engine had to be low in cost. A target price of $400 had been set, about half the cost of any competitive engine. Also, if a low cost was to be achieved, it had to be simple and rugged with as few parts as possible.

Early in the planning process, a horizontally opposed four-cylinder layout was selected, with the cylinders cast in pairs. It afforded several technical advantages over the radial design it proposed to replace. It offered better forward visibility, and allowed for better aerodynamic streamlining. A simple two-throw crankshaft was used, supported by two sturdy lead-bronze main bearings, with no center main. The one piece cast aluminum crankcase was fairly intricate, but well within the capabilities of many Detroit foundries.

Finally, an L-head valve arrangement was selected, which, although somewhat inefficient, was simple and basic. To save weight, the valve mechanism was left uncovered, even though this required the valves stems to be manually lubricated before flight. To further reduce costs, it appears that some parts were borrowed from the industrial engine line then being produced by the company. The choices made by the design team were certainly conservative and reasonable. It is not known how much Insley had to do with this project, but certainly Harold Morehouse had none; he was not hired until 1933.

By midsummer of 1930, the experimental shop craftsmen had completed a prototype model and dynamometer testing began. In order to test the motor in flight, and thus qualify for an Approved Type Certificate, the company purchased a little single seat Buhl Pup monoplane from its Michigan neighbor, the Buhl Aircraft Company, of nearby Marysville.

A Buhl Pup monoplane, similar to the one used to test the first A-40. This plane has a Szekely motor intalled. JUPTNER

CONTINENTAL A40 ENGINE

ANNOUNCEMENT has been made by the Continental Aircraft Engine Company of Detroit that its new Model A40 light airplane engine is now in production. The Continental Model A40 engine is a four-cylinder, L-head, four-cycle design of the horizontally opposed type. The engine has 3 1/8-inch bore and 3 3/4-inch stroke, which gives a piston displacement of 115 cubic inches. It develops 35 horsepower at 2,500 revolutions per minute and 38 horsepower at 2,800 revolutions per minute. The engine is air-cooled and weighs 138 pounds with magneto and carburetor.

The cylinders are cast in pairs of nickel iron. Cooling fins are cast on the cylinders. Provision is made for circulation of air around each barrel. The bores are ground to a mirror finish and held to close limits. Each cylinder block is attached to the crankcase by six studs and nuts.

The cylinder heads are cast in pairs of heat-treated aluminum alloy. Adequate fins for heat dissipation are cast on the heads, and bronze spark plug inserts are cast in place. The cylinder heads are removable and attached to the cylinder blocks with studs and nuts. Steelbestos gaskets are used between head and block.

Trunk type pistons are made from heat-treated, permanent-mold aluminum alloy castings. The heads are thick to dissipate heat, and the skirts are extra long to eliminate scuffing. Three compression rings and one oil control ring are above the piston pin boss. Piston pins are secured by snap rings.

The crankcase is a one-piece, heat-treated aluminum alloy casting heavily ribbed within for strength, and presenting a smooth exterior which conforms very well to the airplane's nose cowling. The case contains crankshaft and camshaft bearings, and carries the valve tappets in reamed bosses. An oil reservoir is provided in the base of the crankcase. This reservoir is divided from the upper portion by wide webs between the tappet bosses which extend toward the center a considerable distance from each side wall, and between them is inserted an oil screen which runs the length of the case. An oil filler and breather is provided on the forward part of the case, while on the rear face are cast four supporting lugs.

A crankcase cover of heat-treated aluminum alloy is attached to the rear of the crankcase and carries the flange mounted magneto which is driven from the rear of the crankshaft. The crankshaft rear bearing support is also bolted to the crankcase cover. On the lower part of the cover is the tachometer connection.

The crankshaft is a one-piece, three-throw high carbon steel forging. It is heat-treated and completely machined. Counterweights are forged integral with the crankshaft. The shaft is drilled throughout for lightness and plugged to provide an oil passage. It is supported by two bronze-backed, babbitt-lined bearings which are align-reamed in place. Thrust is taken in either direction to provide for tractor and pusher installations. The crankshaft is machined with an S. A. E. taper end for propeller hub, which is furnished with the engine. The rotation of the crankshaft is anti-clockwise when viewed from the propeller end.

The camshaft is a heat-treated carbon steel forging drilled throughout for lightness and to provide an oil passage. Individual cam lobes and bearing journals are case hardened and ground. A drive is provided at the rear end for tachometer. The forward end is slotted to operate the oil pump blade.

The connecting rods are made from heat-treated carbon steel one-piece forgings. The caps are severed from the rods and babbitt crankpin bearings are permanently spun in after the caps have been bolted to the rods. Bronze piston pin bushings are pressed into bosses and are reamed in place.

Chrome nickel, tungsten steel valves are operated by long, round-nose steel tappets. Valve adjustment is provided by screws and lock nuts.

Fuel mixture is provided by a single carburetor attached to the bottom of the crankcase. The riser from the carburetor passes upward inside the case, where it branches right and left and connects with the cylinder blocks by means of intake pipes. The oil serves to heat the intake charge, and the cold mixture cools the oil.

Full pressure lubrication is supplied to main bearings, crankpin bearings and camshaft bearings. An automatically primed blade type oil pump is operated by the camshaft, and oil is forced through drilled passages in crankcase, crankshaft and camshaft to the various bearings. Intake of pump is through a drilled boss extending to the bottom of the oil reservoir. Cylinder walls and piston pin bearings are lubricated by oil spray. No internal or external oil tubes are used in the engine. The oil reservoir holds three quarts of lubricant, which is measured by a bayonet oil gauge toward the front of the case. A deep annular groove in the rear camshaft journal, provided with a drilled hole into the camshaft oil passage, registers with an opening in the crankcase where the pressure relief valve is located. Oil pressure is maintained at approximately 35 pounds at 2,500 revolutions per minute.

Ignition is provided by a single magneto driven directly from the crankshaft. A tachometer drive is provided.

Specifications

Bore	3 1/8 inches (79.4 mm.)
Stroke	3 3/4 inches (95.2 mm.)
Piston displacement per cyl.	28.7 cubic in.
Total piston displacement	114.8 cubic in.
Power	35 h.p. at 2,500 r.p.m.
	38 h.p. at 2,800 r.p.m.
Dry weight	135 pounds

This February, 1931, announcement contains some estimated specifications that were later changed. The horsepower rating was revised to 37 HP at 2550 RPM, and the weight increased to 147 pounds which included the propeller hub.

AERO DIGEST

CONTINENTAL A-40

Continental Aircraft Engine Company, Detroit, Michigan

• President and General Manager: W. R. Angell. Vice-president: Roger Sherman. Sales Manager: Ray A. Long. Chief Engineer: N. N. Tilley.

Four-cylinder, horizontally opposed, direct-drive, air-cooled. Rated by the manufacturer at 40 h.p.; by the Department of Commerce at 37 h.p. at 2550 r.p.m. ATC 72. Wet sump lubrication, pressure feed to crankshaft and camshaft bearings. Compression ratio 5.2:1. Bore 3.125 in. Stroke 3.75 in. Displacement 115 cu. in. Length overall 27.31 in. Height overall 19.44 in. Width overall 28 in. Bare weight (with magneto and carburetor) 144 lbs. Weight per h.p. at sea level 3.9 lbs. Fuel consumption .75 lb./b.h.p./hr. Oil consumption .025 lb./b.h.p./hr.

Construction: Crankshaft steel forging machined all over, one piece, three-throw. Crankcase cast aluminum alloy, one piece, heavily ribbed within for strength. Plain bearings. Pistons cast aluminum alloy, trunk type, three compression, one oil ring above gudgeon pin, gudgeon pins floating, located by aluminum buttons. Cylinder heads cast aluminum alloy, L-type, attached to cylinder by nickel alloy steel studs and nuts. Cylinder barrels cast iron, each block attached to crankcase by six studs and nuts. One intake and one exhaust valve per cylinder.

Accessories: Scintilla PN4D, F1 or Bosch FF4A magneto, Stromberg NAS2 carburetor, B. G. 4B2 spark plugs.

A-40 specifications — AERO DIGEST

The Pup's 26 HP Aeronca engine was replaced with the Continental A-40, and on November 10, 1930, the first flight was made. (This plane was Buhl's prototype Pup which was superseded by the production model using a Szekely motor).

Paul Wilcox, a grandnephew of Angell, the company president, had been hired to do the test flying. By the spring of 1931 he had put on the required fifty flight hours; the Continental A-40 received its Aircraft Engine ATC #72 on May 10, 1931. The announcement to the aircraft trade is reproduced on the preceding page. However, the development process, both before and after certification, was to prove costly and difficult.

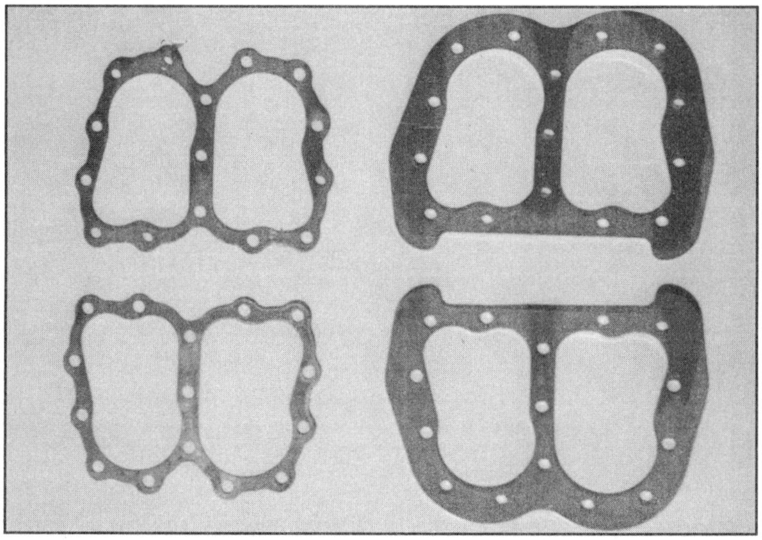

Gaskets- thick soft aluminum sandwitched between two layers of thin copper. — AUTHOR

Following its February, 1931 announcement, a number of light aircraft manufacturers showed an interest in the new A-40 and several bought prototype models for test installations in their airframes. The Taylor Airplane Co., of Bradford, Pennsylvania, was one of the first. C. G. Taylor, the company's president, had completed his first E-2 Cub, but had no really suitable engine. The prototype was powered by a French Salmson 9-cylinder radial, which worked well. But it was too expensive, and its availability from foreign suppliers was questionable. When Taylor heard that Continental Motors in Detroit were making a new small aircraft engine, he and his partner, Bill Piper, drove up to inquire. They were impressed by what they saw, and since they knew Continental to be a reliable company, they bought one on the spot, engine # 132.

Back at the Cub factory, the new motor was quickly installed and test flights begun. Almost immediately, a number of difficulties were experienced, all related to overheating. The head gaskets, made of asbestos and steel would not allow the heat from the cylinders to be transmitted to the aluminum heads, and thus dissipated into the air. When the heat built up, the gaskets would blow. The Continental test pilot, Wilcox, was well aware of this problem, he had reported numerous forced landings due to blown gaskets during the winter of 1930-31. Finally a solution came, not from company research, but from an inventive-minded customer. Paul Wilcox recalled:

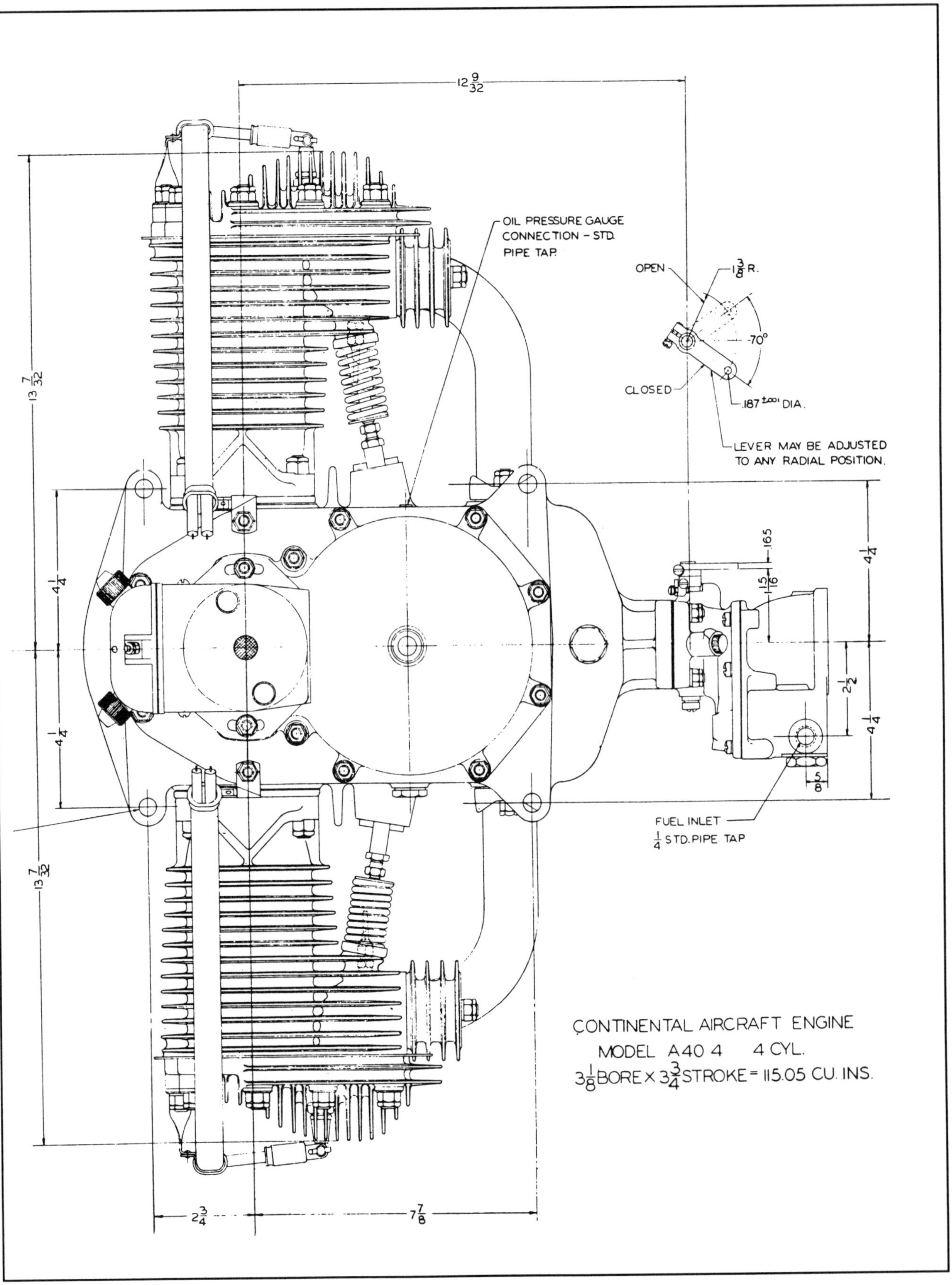

OIL PRESSURE GAUGE CONNECTION - STD. PIPE TAP.
OPEN
CLOSED
70°
.187 DIA.
LEVER MAY BE ADJUSTED TO ANY RADIAL POSITION.
FUEL INLET 1/4 STD. PIPE TAP
CONTINENTAL AIRCRAFT ENGINE
MODEL A40 4 4 CYL.
3 1/8 BORE X 3 3/4 STROKE = 115.05 CU. INS.

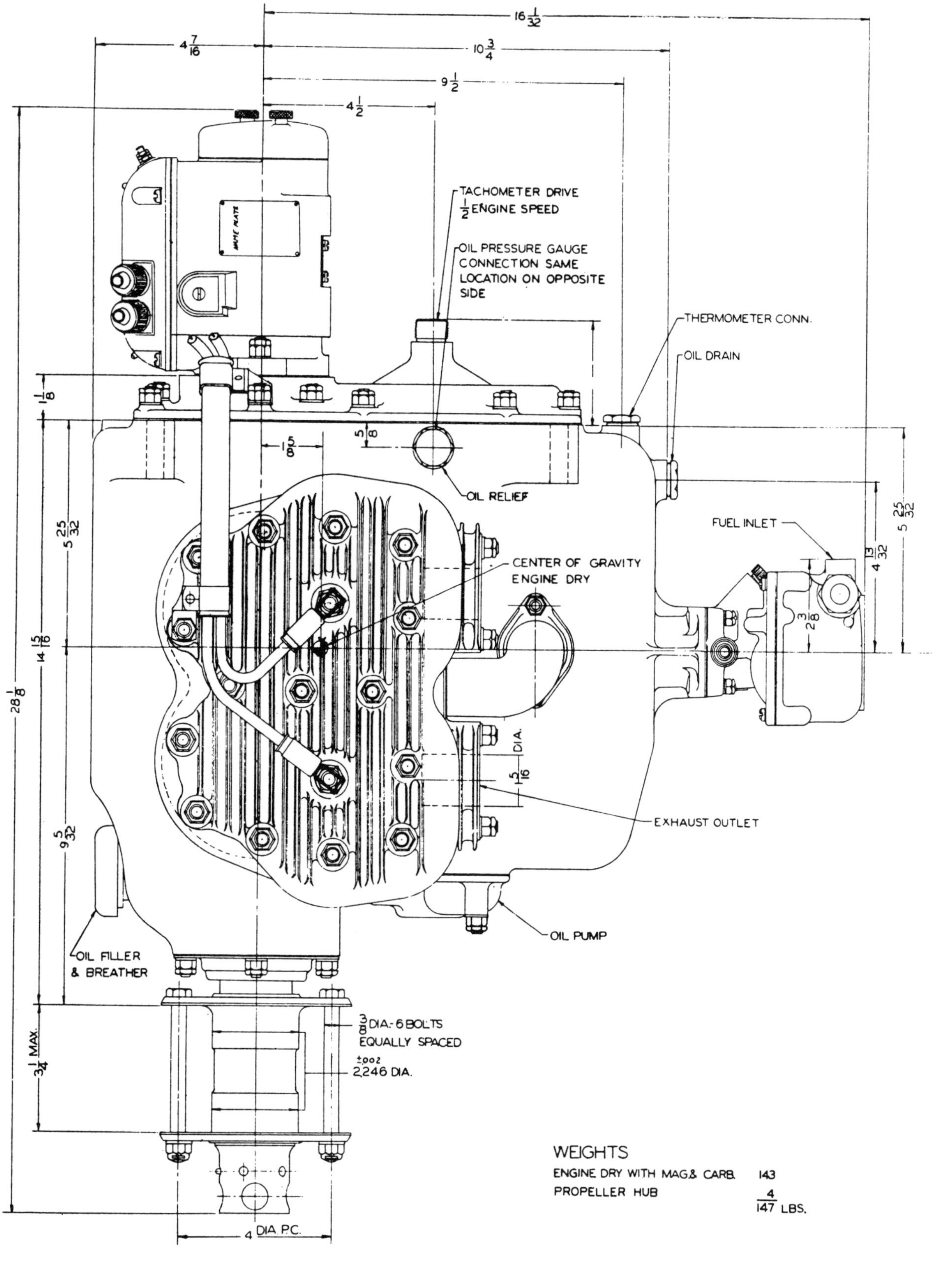

WEIGHTS

ENGINE DRY WITH MAG & CARB.	143
PROPELLER HUB	4
	147 LBS.

"The company told me about a customer in upper Michigan who was having cooling problems with his A-40, so I flew up there to see if I could help. The young fellow was trying to build a helicopter and had installed one of our new engines. By the time I got there he didn't need any help. He showed me how he sandwiched a thick sheet of aluminum between two sheets of thin copper to make a head gasket. This permitted the heat from the cast iron cylinders to come up to the heads where it was dissipated.

"I couldn't get back to the factory fast enough and report to our engineers. We hand made some gaskets which I installed on the engine in our little test plane. I abused the engine all I could by climbing at a steep angle and flying at low speeds, thus heating up the heads. The new gaskets held and from that time on we had very little trouble."

Ole Fahlin testing one of his props on an early A-40 JUPTNER

Another difficulty, the burning of exhaust valves, was of course related to the overheating of the cylinders. The new head gaskets improved this situation somewhat, but more cooling of the exhaust area was needed. Wilcox suggested adding a finned aluminum exhaust fitting:

"With the help of a draftsman, I designed aluminum fins for the exhaust ports which helped cool the valves and cured most of the valve trouble. Along with this we installed a simple O-ring attachment where the intake pipe went into the crankcase."

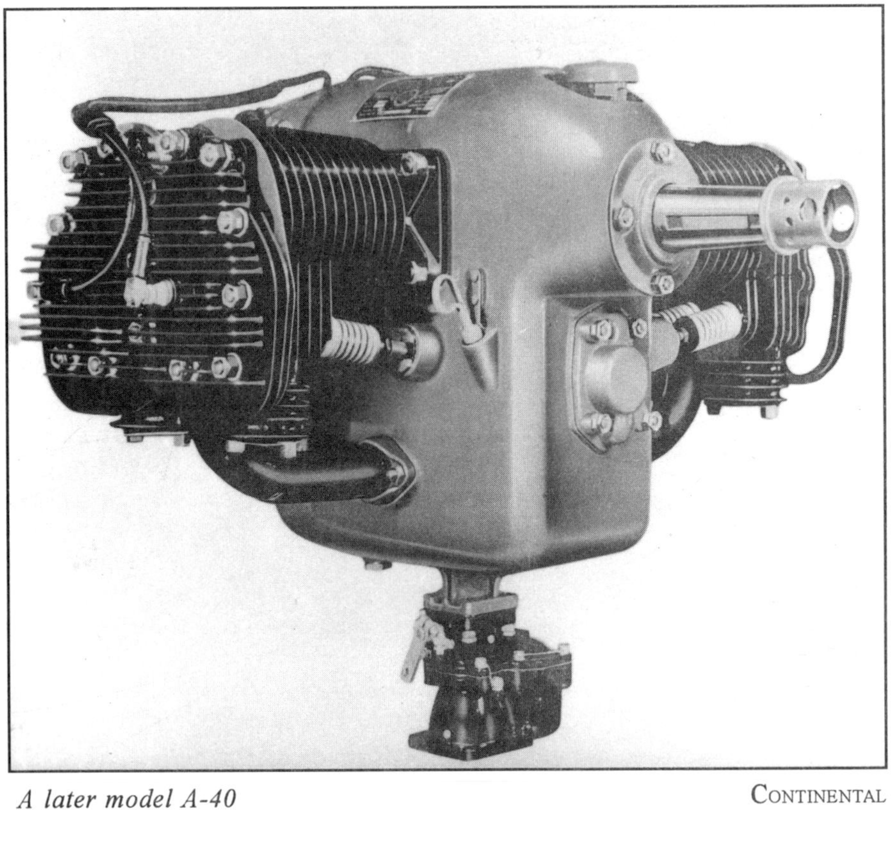

A later model A-40 CONTINENTAL

Finally, the magneto failures were solved by the replacement of the Bosch model FF-4A with the Scintilla PN4-PN1, which in turn was replaced by the Bendix Scintilla SF-4L. Of course when the heating problems were solved, it eliminated many of the magneto problems as well.

By comparing the prototype rendering shown at the opening of this chapter with photos of later models, a number of improvements may be noted. The Cylinder heads have larger cooling fins, the cylinders have deeper cast fins, and the crankcase appears to have been strengthened. The Zenith Carburetor has been replaced with a Stromberg NA-S2.

This new small Continental engine was accepted with enthusiasm by the light plane industry. It had all the attributes essential in those harsh economic times. It was relatively inexpensive. Aircraft manufactures paid approximately $400 for each engine, a far cry from the two or three thousand dollar cost of the big radials. Continental was a reliable company, its name was respected throughout the country. The engine's warranty was reasonable, 50 hours or three months. This was scrupulously honored, and often extended when recurrent troubles arose.

The A-40 offered several advantages to those manufacturers attempting to market small planes. As noted earlier, the initial cost was low, at least $200-$300 below any competitive engine. Operating costs were minimal. Fuel consumption was a miserly 2.8 gallons per hour. While not recommended, any auto fuel with at least a 72 octane rating could be used. Oil consumption was a quart in five or six hours. A top overhaul kit, needed every 200 hours or so, cost less than $10. Major overhaul parts cost a modest $20. It was a simple design; most repairs could be done by a reasonably skilled mechanic with ordinary hand tools.

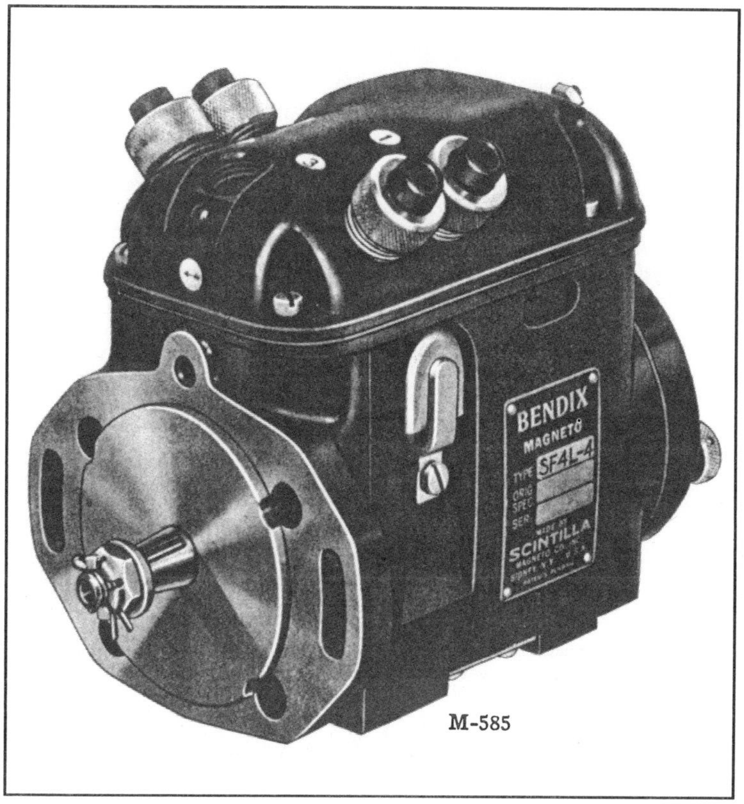

Bendix Magneto

The Bendix SF4L magneto provided ignition for almost all the A-40's. Early in the development process, as mentioned earlier, it had replaced the Bosch FF-4A and the Scintilla PN4-PN1. It was a more modern design, with no separate distributor blocks; each wire entering the top of the case through radio shielded connections. Maintenance was simple, just remove the back cover during routine inspections, clean out any grease or dirt, and oil the felt wick that lubricated the cam. Overhaul was recommended at 400 hr intervals. At that time the two ball bearings should be greased and the contact points replaced, an inexpensive procedure.

"Single Ignition" was often considered a red flag by pilots accustomed to the almost universal practice of using dual magnetos on aircraft engines. However, in actual practice, the Bendix magneto seldom failed. Forced landings were more often caused by sticking valves, dirty gasoline, or other examples of poor maintenance. The single ignition system was cheaper, both in initial cost and maintenance. Although Continental later made a dual ignition model available, (the A-40-5), their official policy was "Dual ignition is unnecessary except in foreign countries where it is mandated by law"

The author and wife Marian ready to go aloft in their A-40 powered E-2 Cub. AUTHOR

After first trying a Zenith automotive type carburetor, Continental selected a Stromberg NA-S2 as its standard for production. It was a very basic design with few field adjustments necessary. The main metering jet was fixed, while the idle jet had a simple needle valve adjustment which regulated the richness of the mixture at idling speeds. This idling adjustment also allowed the engine to be self-primed, since the jet opened in the carburetor throat above the closed throttle butterfly valve. Except in very cold weather, no external priming system was needed.

A heated air muff was usually unnecessary, because the carburetor bolted directly onto the hot aluminum crankcase. The fuel-air mixture passed through this crankcase also, reducing the possibility of carburetor ice. Given reasonable maintenance, and using clean fuel, engine failure due to carburetor malfunction was rare when flying an A-40.

Over the next several years, a number of planes received certification using the Continental A-40. The first was the Taylor Cub which received Group 2 Approval #2-358 on 6-11-31. Next the Alexander Flyabout received ATC #439 on 7-31-31. The stories of all the A-40 planes will be told in later chapters. Only two models attained any volume production using the A-40, the Cub and the Taylorcraft. Taylor, (later Piper), built about 1800 of their E-2, J-2 and J-3 40 HP models. Taylor made 600 of his sleek little Model A Taylorcrafts.

Two types of A-40 crankshafts. The early, fully machined, type on the left, the forged type on the right AUTHOR

An A-40-5 installed in a Taylorcraft Model A AUTHOR

The design of the A-40 was basically sound, and few, if any, bulletins were ever issued on it. The early troubles, magnetos and head gaskets, were quickly solved. One serious defect appeared later. The crankshaft would break at the rear rod journal at almost the exact 100 hour mark! Bill Jones, a retired FAA inspector once recalled:.

"We continually broke crankshafts on our early A-40's. But we were near Detroit, so we would just load up the little engine and take it to Continental's service shop. In less than three hours, they would have a new crankshaft installed".

Walter Jamouneau mentioned this problem during an interview with the Author some years ago. He didn't remember the cause, but he did say it was one of the reasons Taylor tried to design his own engine. Some suspected it was due to the lack of a thrust bearing, but a careful examination of early engines does not support this theory. Early crankshafts were machined out of a solid billet, and had a counterbalance, perhaps something in the manufacturing process caused the defect. The later forged crankshafts did not seem to have this trouble.

Few changes were made during the entire production run of the motor. The ignition timing was revised after S/N #371, the magneto was set to fire at 22 degrees instead of 26 degrees. A new camshaft changed the valve timing on the -3 and -4 models. Early A-40's had babbitt lined connecting rod bearings which limited their rated RPM to 2550 top. Later, following the current automotive practice, steel backed replaceable insert bearings were used, which allowed for a 2575 cruise RPM. This was supposed to produce a full 40 rated horsepower. (In actual use, when a Sensenich 69C23 propeller was installed, most Cub owners would be able to run their engines up to 2700 RPM for short periods of time). The factory operators manual lists 2680 RPM as maximum for the -2 and -3 models, 2700 for the -4 and -5's

Due to the fact that the little engines were almost always run at full throttle, some operators had problems with upper cylinder lubrication, resulting in stuck rings and scored cylinders. With a deep bore, (3 3/4"), and small diameter, (3 1/8"), the splash system did not furnish lubricant to the upper cylinder walls in hot weather, even when the recommended 20 weight oil was used.

To remedy this situation, as the last of the A-40's were being produced in 1938, a new connecting rod oiling system was devised. Small holes were drilled in the rods and bearing caps so as to direct oil under pressure into the cylinder bore. This improved lubrication and also aided in cylinder cooling. Best of all, on 3-15-39, Continental offered a retrofit kit so that all A-40-4 and -5 motors could have this system installed for the grand sum of $13.32 plus labor!

Late in 1936, a twin ignition model, the A-40-5 was introduced due to export requirements, mainly those in England. The extra magneto did not increase the available horsepower, but increased the weight by 10 lbs. In fact when checking the mags on these engines there is never any RPM drop. Whether the added safety factor of twin ignition was worth the $125 additional cost is debatable. Due to this weight and cost increase, few were sold in the U. S.

It seems unlikely that Continental ever reached a production volume necessary to show a profit on the A-40. If all the known production of planes is added together, the number appears to be near 2500. Allowing for certain engines to be sold as replacements, the total production might be estimated at 2800-3000. The A-40 serial numbers started at #100. The -2 series ran to about 450, according to notes in the parts book. The -3 and -4 numbers appear to be mixed together, but all serial numbers above 1000 appear to be -4's. The highest number of a -4 that the Author has noted was #2538. The A-40-5 twin ignition models started with serial 3000, probably less than 200 were built.

Using these records to estimate production, 3000 seems to be a reasonable figure. The revenue per engine was about $400, which would suggest a total sales volume for this engine of $1,200,000. If $300,000 parts sales are added the total could have reached 1 1/2 million dollars, not a huge amount even during the depression.

The next chapter will cover the operation and maintenance of the A-40.

Jeannie Hill enjoys "Flying with 40 Horses" JEANIE HILL

Specifications
Continental A-40
ATC #72, 5-19-31

<u>MODELS</u>: Continental A-40, A-40-2, A-40-3, A-40-4

<u>T.C. NUMBER</u>: ATC 72

Model	A40-4	A-40, -2, -3
Type	4HOA	--
Rating:		
Maximum continuous, hp, rpm, at S.L. press. alt.	40-2575-S.L. 38-2500-S.L.	37-2550-S.L. --
Take-off (one minute), hp, rpm, at full throttle	40-2575 39-2500	37-2550 --
Fuel (minimum octane aviation gasoline)	73	--
Bore and stroke, in.	3.125 x 3.75	--
Displacement, cu. in.	115	--
Compression ratio	5.2:1	--
Weight (dry), lbs.	144	--
Crankshaft gearing	None	--
Propeller shaft, SAE No.	0	--
Supercharging (Impeller gear ratio)	None	--
Carburetion	Stromberg NA-82 carburetor	--
(Venturi diameter, in.)	1-1/8	1-1/8 or 15/16
Ignition, single	Scintilla PM4-DF, Bosch FF4AC or Bendix-Scintilla SF-4L	--
Certification basis	Approved Type Certificate No. 72	

NOTE 1. Maximum permissible cylinder head, barrel and oil inlet temperatures, 550 degrees F., 300 degrees F., and 200 degrees F., respectively.

NOTE 2. Approval expired 11/1/41. No engines of these models manufactured after this date or with serial numbers above 2699 eligible for use in certificated aircraft.

Specifications
Continental A-40-5
ATC #174, 2-15-36

<u>MODELS</u>: A-40-5

<u>T.C. NUMBER</u>: ATC 174

Model	A-40-5
Type	4HOA
Rating:	
Maximum continuous, hp, rpm, at S.L. press. alt.	40-2575-S.L.
Take-off (one minute), hp, rpm, at full throttle	40-2575
Fuel (minimum octane aviation gasoline)	73
Bore and stroke, in.	3.125 x 3.75
Displacement, cu. in.	115
Compression ratio	5.2:1
Weight (dry), lbs.	156
Crankshaft gearing	None
Propeller shaft, SAE No.	0
Supercharging (Impeller gear ratio)	None
Carburetion	Stromberg NA-82 carburetor with 1-1/8 in. venturi
Ignition, dual	Bendix-Scintilla SF-4R magnetos
Certification basis	Approved Type Certificate No. 174

NOTE 1. Maximum permissible cylinder head, barrel and oil inlet temperatures, 550 degrees F., 300 degrees F., and 200 degrees F., respectively.

NOTE 2. Approval expired 11/1/41. No engines of this model manufactured after this date or with serial number above 3235 eligible for use in certificated aircraft.

Chapter Three

Care and Feeding

(Fixing and Flying the Continental A-40 Aircraft Engine)

Typical Continental A-40 installation AUTHOR

Disclaimer: Those operating A-40 engines today should use the Continental Operators Manual for accurate operating and maintenance information. The Cub Club has copies. The information below reviews the experience of many "old timers", but should not be considered official.

The A-40 was the perfect answer for the lightplane flier of the 1930's. Inexpensive to buy, it was simple to operate and easy to maintain. Best of all, it was a delight to fly. Smooth and reliable, it allowed thousands of new pilots to taste the delights of flying in those early days. It gave the light aircraft industry a new lease on life.

But, as with all mechanical things, this little engine has a few quirks that should be recognized. As a convenient way to explain its operation, lets take an imaginary flight in an A-40 powered E-2 Cub.

Flying the Continental A-40

First one must recognize that it is a "single ignition" motor meaning it has just one magneto and one set of spark plugs. This might sound like a dangerous situation to some, but in actual practice the failure of a magneto is a remote possibility. The Author has flown his A-40's almost fifty years, and never had a forced landing due to magneto trouble. (Knock on wood). After only a few flights, you will surmount this psychological barrier and forget it has *"only one mag"*.

Preflight inspections are simple, there is not much to check. Is there gas in the tank, and 4 quarts of oil in the sump? One ritual, though, must be observed. Oil the valves. The A-40 valve operating system is completely exposed and must be lubricated manually. The factory specifications called for cleaning with kerosene and oiling with penetrating oil.

These days there are a number of good lubricants available in convenient Aerosol cans. **Always** spray the valve stems before every flight! If you don't, the valve stems will become dry, they will stick, the engine will lose power and a forced landing will follow.

As you check in the cockpit, you will note there is only a bare minimum of instruments and controls. Engine instruments will consist of a tachometer, oil pressure and oil temperature. If you are fortunate enough to be flying a Taylorcraft A or an Aeronca KC, all will be contained in a beautiful six-inch round instrument in the center of the "dashboard". There will be a throttle, of course, and some sort of gas shut-off. No primer and no carburetor heat. Neither is necessary, the carburetor is fastened directly to the bottom of the oil pan, and stays quite warm. Priming is furnished by the NA-S2's idle bypass jet.

Now to starting. Because it is likely that the plane you are flying will have no brakes, use chocks, or better yet, tie the tail down. If you are alone, it is possible to prop the plane from behind, using your foot for a chock, but use caution. First crack the throttle slightly.

It is possible to start the A-40 standing behind the propeller. AUTHOR

A correct stance when starting the A-40 AUTHOR

Next, with the switch off, pull the prop through smartly half a dozen times or until you see a few drops of gasoline dripping from the air intake. Now the engine is ready to start. Turn the switch on, lay your hands on the propeller blade about a foot from the hub, and give it a sharp flip. Remember that the single magneto does not have an impulse, so you must give it good rotational speed to get a strong spark. The engine should start.

After the engine starts, look in the cockpit and see if the oil pressure comes up to about 25 pounds. If it doesn't come up after about 30 seconds, shut the motor down. The A-40 has its oil pump high in the front of the engine, and when installed in a tail wheel plane, as most are, the oil can drain out. Then, if the pump doesn't prime itself, you have no oil pressure, and damage can result. But there is a simple remedy. Pick the tail of the plane up and hold it over your head for a few seconds. The oil will run to the front of the engine and prime the pump. Since any A-40 powered plane would be quite light, this is no great problem.

With the engine now running, untie it and climb into the cockpit. The engine should be idling smoothly at 5-600 RPM. Advance the throttle slowly, and don't be concerned if it won't take much throttle at first. It has quite a large oil sump for its size, and the cold oil may keep the fuel from completely vaporizing, leaving

Aloft in an A-40 powered E-2 Cub. Davis

the mixture too lean. At about 1000 RPM the main jet will take over, and you can get full RPM.

You will be surprised to find that you will need to use about half throttle, perhaps 1500 RPM, to taxi. There is just not much thrust there. When you reach your takeoff point, the engine is probably warm, but to be on the safe side, wait until the oil temperature needle gets off the peg. There is really nothing to check prior to takeoff. No magneto check, you have only one. The lack of brakes will probably rule out a full power check, so just turn into the wind, open the throttle and go. As you start your takeoff roll, glance at the tachometer and check the reading. On most planes, it should be 2200-2300 RPM. As you lift off and gain airspeed, this will rise to 2400+ and you will be flying--with 40 horses!

After you reach pattern altitude and level off, the RPM will probably rise to about 2500 with most planes, depending on the propeller used, of course. You will note that the RPM varies with the airspeed. In a slow, climbing attitude, you may have trouble getting above 2350; in a shallow dive, you may see 2600. Above all, you will notice how delightfully smooth the little engine runs, not nearly the level of vibration of its big brother, the A-65.

As you cruise around the field, you may be startled when the motor "skips" a beat now and then. Some motors will, some will not. Other than causing the pilot to catch his breath, it does not seem to be dangerous, and is probably due to small amounts of carburetor ice. Ted Wells, manager of the Cub plant in the mid-30's, would tell worried customers, "It will spit, but it won't quit."

It is a fact of life that you will fly the A-40 at full throttle more than most engines. Full power is needed for takeoff, of course, and to get any reasonable rate of climb. If two people are in the plane on a hot day, it may take full throttle just to stay in the air. Old timers tell of using full throttle continually when instructing in 40 HP planes. They thought the extra airspeed helped keep the engine cooler. Full throttle creates no real problem unless the oil temperature gets too high, so just relax and enjoy your flight.

In most A-40 powered planes, you won't be going too far. The E-2 Cub will fly at just under 65 MPH, while the J-2 may get up to 70. The Taylorcraft Model A will cruise at 75 with two aboard, perhaps 80 when lightly loaded. The Aeronca KC was one of the fastest A-40 planes, an honest 80 MPH with two aboard. But remember, these planes were really not intended for cross country flying. Their main use was student instruction; short flights around the field.

Returning to land, there is nothing much to do but retard the throttle. You probably will not have a carburetor heat system to apply. When gliding you may be startled by a sudden "bang", as loud as a pistol shot! Don't worry, the little engine tends to backfire with a retarded throttle, and the short exhaust stacks on most planes don't give much protection to the ears.

The airspeed will keep the engine windmilling at 900-1000 RPM, so there is a minimum possibility of it dying. However, an occasional short burst of the throttle might be comforting. After taxiing back to the hangar it is a good idea to let the engine idle a couple of minutes to let it cool evenly. That's about all there is to flying an A-40. Just put in some gas and fly it. All that fun on only 2 1/2 gallons per hour!

Care and Feeding

The little engine only needs two kinds of food, gas and oil. But it is a little finicky about both. It will run fine on 80 octane aviation fuel, and unleaded auto gas. But it will **NOT** run on 100 octane so-called low lead avgas. The compression is not high enough to deal with the lead content. After a few minutes of flight, the plugs will foul up, and down you'll come.

Care must also be taken when selecting the oil to be used. Old timers soon found that the heavy aircraft oil specified for the big air-cooled radials simply would not lubricate the little engine. Bob Thompson, the dean of A-40 mechanics, related this story:

"I very well remember the first A-40-3 I saw. Mr. Ted Wells, Piper's sales manager, had flown an E-2 into the East Dayton airport back in 1934, and was remaining overnight. He ordered an oil change, so I looked in the Standard Oil chart and it said to use Standard Oil 40 weight. I put in a gallon, cowled the little engine back up and went home for the day. Mr. Wells came out the next morning, got a tank of gas and a prop and was on his way. About 15 miles out, the little engine froze up and down came Mr. Wells.

"He called up the airport manager and had him bring out a gallon of 20 weight auto oil and poured it in. He apologized for forgetting to tell me about the little engine's queer whims. The small bore and long stroke make it hard to get oil splashed up to the top of the cylinders."

Honeycomb air inlet pipe AUTHOR

This Taylorcraft factory photo shows an air cooling scoop installed on an A-40 AUTHOR

The factory approved air scoop on an A-40 is fitted with honeycomb vanes, to pick up the air and direct it into the venturi and without turbulence. If someone puts a non-vane air scoop on an A-40 in ignorance, erratic running and some loss of power will result due to uneven fuel distribution.

The A-40 is extremely easy to repair. Let's start with a top overhaul, with parts costing only $7.51 (in 1936). The motor can even be left on the motor mount. After removing the spark plugs and ignition wires, undo the six nuts holding the intake and exhaust manifolds. Remove the manifolds, then take off the six 7/16 nuts holding down each cylinder block. The cylinders can now be pulled off the engine. Push out the wrist pins and the pistons can be removed. With these parts on the workbench, the heads can be removed so that the valves are exposed. New piston rings can be fitted, and the valves reground in a matter of an hour or so.

There is one maintenance item that can be a source of trouble if not handled properly. The aluminum cylinder heads are separated from the cast iron blocks

A-40-5 twin ignition AUTHOR

by a thick (.040) soft aluminum gasket sandwiched between two of very thin copper. This metal-to-metal contact lets the cylinder block heat flow to the deeply finned aluminum heads for dissipation. The gaskets allow for the difference in expansion between the cast iron and aluminum.

When new gaskets are installed, care should be taken not to over-torque the hold-down nuts. The factory manual suggests a rather inexact "solid firm tightness". After the initial run-in, these head nuts should never be retightened. These may seem loose when the engine is cold, but they will tighten up again when hot. The temptation to tighten may be strong, but leave them alone. Continual tightening will only squeeze out the soft aluminum, and they will eventually leak or blow out, causing a forced landing. Always install the pal nuts used by the factory.

When the pistons are removed from the connecting rods, the rods themselves can be removed by unscrewing the two 9/16 inch nuts on their lower end. Some of the earlier rods had babbitt bearings, but most had the replaceable lead-bronze crankpin bushings. They can easily be forced out with finger pressure. In 1936 the total cost of a replacement set was $4.00, so new ones were usually installed. The oil pressure relief valve, a simple spring and ball device, can be taken out, cleaned and reinstalled.

At this time the tappets can be removed by unscrewing the four 5/16" nuts holding down the tappet guide plates. They can then be pulled out and the packing and glands can be removed by inserting a finger in the hole. If you are fortunate enough to have new packings, use them but these chevron gaskets seldom leak. The oil pump can be dismantled, and the front crankshaft oil seal removed, if you have pulled the propeller. Both gaskets are easily replaced but be sure to use a Continental factory gasket on the oil pump. It regulates the clearance of the oil pump blade and the housing, and must be the right thickness.

That would take care of a top overhaul, but at this time if you should decide to do a complete "major", it would require only a few more hours of work. You will need to take the motor off the engine mount and place it on a workbench or engine stand.

It is again a simple matter to remove the magneto, take the back plate off the motor and pull out the rear main bearing support. Then just slide the crankshaft out the back of the case for inspection. Reassembly is equally simple, but be sure to use the proper shims under the rear bearing support to get correct end play on the shaft. Camshaft timing is set by marks on the crankshaft and camshaft gears, so you can't go wrong here.

In 1936, the cost of a complete major overhaul kit was $20.00, but since replacement of the two brass main bearings usually involved some machine shop work, this was often avoided. After hundreds of hours of flying, it was possible for the front brass bearing to wear egg shaped, thus allowing oil to flow out too easily. This usually would be indicated by low oil pressure when the engine was hot.

The famous Bob Thompson running up an overhauled A-40 in his Heath racer AUTHOR

Cylinder banks and the intake tubes are put on in the reverse order in which they were removed. Do not put a gasket under the cylinder barrels, use only a sealer. Heat should transfer easily when there is metal-to-metal contact between the case and the cylinders. And be VERY careful to tighten the six hold-down nuts evenly! If you don't, it is quite possible to pull one stud out and ruin a crankcase. The same "no gasket" rule holds true for the aluminum exhaust manifolds. Metal-to-metal contact is needed there also.

The magneto and carburetor used on the A-40 are as simple and easy to maintain as the engine itself. The Bendix-Scintilla SF-4L needs only to have the points inspected and replaced or trued up as needed. Adjust according to manufacturers specifications. Put a drop of oil on the felt bearing wicks. Inspect and replace any cracked or damaged spark plug wires. Fasten the wires firmly to the plugs and engine; they are out in the slipstream.

The Stromberg NA-S2 carburetor inspection is equally simple. Drain the float bowl of any water or dirt by removing the 1/4 inch pipe plug. Unscrew the fuel straining screen, clean and reinstall.

To repeat, if you intend to do maintenance or repair work on an A-40, you should get a copy of the official Continental's Operators Handbook. The entire instructions on care and operation cover only fifteen typewritten pages.

****

So, given ordinary care in operation and maintenance, the antique Continental A-40 will give you many hours of carefree, enjoyable, inexpensive, flying. And you will have the added satisfaction of cruising behind a truly unique piece of aviation history; the little engine that made possible America's light aircraft industry.

Chapter Four

Taylor E-2 Cub

Taylor factory publicity photo, 1932 TAYLOR

It was a match made in the heavens!

When the Continental A-40 engine was mated to the Taylor E-2 Cub, the success of both was assured. Without the Continental A-40, the Cub would have become one of the dozen or so "also rans", limping along with a Salmson, Szekley or some other unsatisfactory powerplant. But at the same moment that Taylor was designing the first Cub, the Continental motors corporation was designing their first lightplane motor, the A-40. This is how they came together.

Clarence Gilbert, (C.G), Taylor designed the first Cub, and the Taylorcraft that followed it some five years later. He was born September 25, 1898, the first son of Arthur and Clara Taylor, immigrants from England. Arthur Taylor operated a machine shop in Rochester, and in the booming industrial economy that preceded World War I, the business prospered. Both of his sons, Clarence Gilbert and Gordon literally grew up in the machine shop; it was natural that they would have an interest in all things mechanical. At the age of six, standing on a box because his small stature, C. G. could run a lathe or drill press.

While C. G. showed remarkable mechanical skills, at an early age he developed a serious handicap that would plague him his entire life. When only one year old and just learning to walk, his mother Clara noticed he was always falling to his right side. He developed a high fever, and after it had subsided, his right leg was partially paralyzed. He had been stricken with polio and would thereafter be forced to walk with pain and a serious limp. How this handicap was reflected in his obvious drive for accomplishment and recognition, can only be conjecture. His career certainly reflects an uncommon determination to succeed.

Working in the defense industry during WW I, Taylor became interested in aviation, studying the various technical books then available on the subject. Both he and his brother Gordon began to dream of a career in flying.

NORTH STAR
AERIAL SERVICE
CORP.
29 WESTGATE TERRACE
ROCHESTER, N. Y.

North Star Logo TAYLOR

However it took some years to realize this dream. The responsibilities of marriage and a growing family dictated safer and more lucrative work. Finally, in1925, C.G. and his brother were offered an opportunity to enter the aviation field. A Mr. Jack Loomis of Binghamton, New York, had advertised a WWI Curtiss Jenny for sale. The price was $750, and he would teach the buyer to fly as part of the deal. The Taylors pooled their savings, and borrowing some from their father, bought the plane. With C. G. as a passenger, Loomis flew the old ship to the Taylor family home near Rochester where it was dismantled for a complete rebuild.

This was C. G. Taylor's first opportunity to use the aircraft engineering knowledge he had obtained by studying at home, often at night, the few books then available on aircraft design. He intended not only to overhaul the plane, but to improve and modify it as well.

The brothers took the fuselage completely apart, replacing the dry-rotted lower longerons. At the same time they redesigned the fuselage by adding an additional cockpit. This allowed the plane to carry two paying passengers instead of one, effectively doubling the revenue for each flight. The nose radiator was replaced with a Standard J-l vertical radiator, allowing the front cowling to be streamlined. These changes would have moved the center of gravity substantially to the rear, and it appears the 16 inch stagger was reduced somewhat to maintain a correct balance.

When the plane was reassembled, Loomis, as agreed, taught C. G. to fly over a period of about two weeks, although he was highly skeptical of the "modifications". The main instruction method was "*Follow me on the stick*". After this short instruction period, C. G. was "soloed" and proceeded to teach his bother Gordon to fly.

In the Spring of 1926, they formed the impressive-sounding "North Star Aerial Service Corp." and began barnstorming around the Rochester and Sodus, New York area. Business was good; profits ran as high as $400 a day on occasion. They hauled passengers, took photographers aloft for aerial news photos, and put on exhibition flights.

Their next aviation opportunity came from their instructor, Jack Loomis. As he watched the performance of their remodeled Jenny, he became impressed with the design ability of C. G. Taylor. In the Fall of 1926, a deal was struck whereby the Taylors would build Loomis a completely new plane to use for passenger flights.

Taylor's three-cockpit Jenny TAYLOR

Loomis Special, barnstsorming near Rochester NY TAYLOR

The resulting "Loomis Special" was a high wing monoplane of quite modern design which reflected the principles and ideas C. G. had developed over the years. It had a Warren truss welded steel fuselage; no wires or turnbuckles were needed to keep it in rig. The wide split axle landing gear was much advanced over the spreader bar used on most planes of that era. The wing structure consisted of spruce plank-type spars, the ribs were built-up spruce trusses. This wing design would later be used on the Chummy, Cub and Taylorcraft. The choice of a Clark-Y airfoil enabled the Loomis plane to lift a heavy load at low speeds with a minimum of power. The characteristics were quite similar to those of the USA 35-B used later on all the Cubs.

It had a two to three passenger cabin, and a large open cockpit behind which could seat two with dual controls. Thus the plane could be used for both training and passenger hops. It is interesting to note that the door of this first Taylor-manufactured plane was very similar to the Taylorcraft doors made 10 years later.

The engine was a French-built 10 cylinder Anzani double-row radial which was supposed to put out 120 HP. Weighing only 320 lbs, and turning a slow 1550 RPM, it was over 200 pounds lighter and had 30 percent more power than the commonly-used Curtiss OX-5.

The Loomis plane, given identification number 2594, Taylor S/N 2, Model A-4, served well in Binghamton and the surrounding area for several years. Records show that it was damaged and rebuilt in late 1927, a Hallet 135 HP engine replacing the Anzani.

The success of the Loomis plane, and the aviation "Boom" following Lindbergh's flight in May of 1927, encouraged the Taylor brothers to enter aircraft manufacturing on a full-time basis. They moved their operations to a garage building at 127 Alexander street in Rochester. In October, the local newspaper gave this account:

"The Taylor Brothers are now manufacturing their Arrowing "Chummy" two-place monoplanes in Rochester, after building and selling a 4-place model last Spring. Two planes are now under construction and will be finished in a few weeks, if the motor, ordered from France, arrives in time. The Taylor brothers claim they have the only two-place side by side plane on the market that sells for $2300."

It was actually February of 1928 before the first Chummy was ready for its test flight. The plane had been towed to the new airfield being built on Scottsville Road behind the family Essex, then assembled outdoors without shelter, the new hangars still being unfinished. A local newspaper reporter described the event:

First Chummy TAYLOR

"With Gordon Taylor, co-builder, as pilot, the small plane ran over the ground 300 feet before it gracefully lifted and began a rapid and steady climb until reaching an altitude of about 600 feet. At that level it circled the field a number of times before returning to the ground.

"Yesterday's success was the realization of the Taylor brother's dream to some day build a plane which would give the aeronautical industry a worthwhile Rochester-made plane.

"It is the hope of the brothers to form an airplane manufacturing company in Rochester, backed by Rochester capital."

Gordon and C. G. Taylor in front of the first Chummy TAYLOR

This first Chummy, given the identification number X-4903, Taylor S/N 3, Model B-2, was a high-wing, strut braced monoplane, a design that would be common to all later Taylor products. The length was 24', the wing span was 34', and with a chord of 5' 3", the wing area was 175 square feet. The cockpit was 38 inches wide, and equipped with two sticks and two sets of rudder pedals, the right seat controls being removable. Surprisingly, the landing gear is of the old straight-axle type, differing from the earlier-built Loomis plane.

By April, the Taylor brothers had formed a new company, the Taylor Aircraft Corporation. Modest local financing had allowed the production of a second Chummy, S/N 4 X-4901 which was to be exhibited at the big Detroit Aircraft Show. It was much improved over their first model, it had a spit-axle landing gear and a Seimens-Halske "Yankee 7" motor was enclosed in a streamline cowling. This appearance at the Detroit show was expected to give the fledgling Taylor Brother's Corporation an opportunity to gain a nationwide market for its new plane. Reports were that it was well received, its modern features, such as the welded steel fuselage, monoplane design and side-by-side seating, set it apart from many competitors.

The second Chummy, X-4901 TAYLOR

But as the brothers would tragically learn, flying was still a dangerous occupation. After the downtown show ended, the Taylor plane was taken to the nearby Detroit airport for flight demonstrations. It was hoped that a number of sales could be made and perhaps a national sales organization established. Gordon had arranged to take a number of "prospects" up for demonstration flights; expecting, of course, that the little plane's performance would impress the passengers.

Something went terribly wrong on one of the first flights. News accounts reported the plane took off, climbed to about 100 feet altitude, banked into the wind, then spun to the ground. Both Gordon and his passenger were killed! Clarence Gilbert, who had been watching the flight, was shocked and devastated. He had not only lost a beloved brother, his dreams of building his own aircraft company may have been shattered as well!

To quiet speculation as to the cause of the crash Gilbert gave a lengthy explanation of the accident in a newspaper interview the next day. He steadfastly asserted that his brother had not been careless, and that the probable cause was the passenger "freezing on the controls". This was often suggested in the case of flying accidents in those days, when the actual reason, human or mechanical, could not be determined.

TAYLOR, LOCAL FLIER, KILLED IN PLANE FALL

DETROIT, Apr. 24.—(AP)—Gordon A. Taylor, twenty-five, of the Taylor Aircraft Corporation, Rochester, N. Y., pilot, and Aaron Rosenbleett of New York City, a passenger, were killed this afternoon at the Ford Airport when a plane Taylor was demonstrating went into a nose dive at an altitude of about 1,000 feet and crashed to the ground.

The plane was a Taylor Chummy monoplane which had been exhibited at the aircraft show last week and removed to the airport for demonstrations.

The accident today which occurred about 1:45 p. m. was witnessed by Gilbert Taylor, Gordon's brother and business partner, who was watching the flight from the field.

Rosenbleett was dead when airport attendants reached the fallen plane. Taylor died a few minutes later as he was being rushed to a hospital. The plane was demolished.

Rosenbleett, it was said at the airport, had come to Detroit in the hope of obtaining a position as a salesman for the Taylor Brothers.

Witnesses said the plane made a good takeoff and circled the landing field once. At an altitude of about 1,000 feet it was seen to bank into the wind and then go into a nose dive.

Brothers Planned To Market New Plane.

Gordon Taylor left Rochester for Detroit All-American Aircraft Show the morning of April 12 in the Taylor Arro-Wing Chummy two place monoplane with which he fell.

The show ended last week, but the Rochester pilot remained in Detroit to demonstrate his ship to prospective buyers.

He was accompanied by George Emeric of Greece, a mechanic in the little factory at No. 42 Allen Street.

Success of their plane assured after more than a year of experimenting and hard work, the two brothers incorporated Taylor Brothers Aircraft Corporation and were planning to expand immediately.

They were about ready to move into larger quarters and expected to produce at least twenty-five planes this year.

Gordon gave up all other work a year ago to devote himself exclusively to the Taylor plane, several of which already have been sold.

The young man was a graduate of West High School and a highly skilled mechanic. His interest in aviation dates from his early boyhood and experimenting with model airplanes.

He lived in Lettington Road, Coldwater, with his parents, Mr. and Mrs. Arthur Taylor.

Besides his parents and brother, he leaves four sisters, Miss Ruth Taylor, a student at the University of Michigan, Ann Arbor; Mrs. Gladys La Due and Mrs. Welpton, both of Rochester, and Mrs. Marjorie Peterson of Newark, N. J.

The family lived at Newark for a time.

Gordon's tragic death ROCHESTER DEMOCRAT

In the face of all these difficulties, the loss of his brother and the crash of his aircraft, it must have taken great determination for C. G. to announce, *"We cannot very well discontinue the work for which Gordon gave his life. My Father and I, and others associated with us, will carry on the manufacture of the "Chummy".*

Manufacturing continued on a limited basis in the Taylor Brother's modest Rochester factory. At least one more Chummy was built that summer, the first Kinner-powered model B-2, S/N 7. An application for Identification Mark, (7303), was made on 8-1-28, with the notation, *"Will be used for demonstration then delivered to purchaser shortly. This ship built in conformity to the stress analysis made by Aerotec Inc. and has a factor of safety of 8-5-3 respectively."* As will be noted on the following pages, this same plane was later remodelled to be used in the Guggenheim competition.

Despite the publicity generated by Taylor's new company, few actual orders were received. However, due to the *"Lindbergh Boom"*, every city of any size in this country wanted to *"cash in"* on the new aviation craze. Representatives of the Bradford, PA Board of Commerce, headed by J. M. Brooder, visited the modest Taylor factory in Rochester during the Fall of 1928, hoping to find a firm that could expand their city's industrial base. Already a thriving small city, with an economy based primarily on the oil and refining business, the city fathers were eager to become associated with the new, and extremely popular, aviation industry.

The Taylor Brothers Corporation was sorely in need of assistance. It appears that after the tragic crash that took Gordon's life, further financing was nearly impossible to obtain in Rochester. Certainly, this safety record would not inspire confidence. The manufacturing facilities were inadequate, small and crowded, unsuited for the production and assembly of aircraft. The planes had to be built and assembled, then dismantled and trucked nearly seven miles to the Rochester airport where they were reassembled and test flown.

Bradford investor's group. Taylor, kneeeling 2nd right, Piper standing right TAYLOR

So the Bradford delegation was welcomed with open arms when they arrived for their inspection of the Taylor Brothers Company. And the Bradford people were favorably impressed with the Taylor organization. The Chummy appeared to be well built, and the Taylor's engineering and mechanical abilities adequate. Brooder later reported that the Taylors possessed "a good character", while the Rochester bankers said the company's financial responsibility was "first class", but that "they have no money".

In discussions with the Bradford group, C. G. Taylor indicated it would take $50,000 to set up a manufacturing plant capable of building 200 airplanes annually. (He would later, in 1936, tell the Alliance, Ohio, Chamber of Commerce that it would take $50,000 to set up the Taylorcraft plant in their city.) If they could raise this amount in a stock sale, he would move to Bradford.

When the delegation returned to Bradford, they quickly obtained the approval from the bankers and oil men on the Board of Commerce for the acceptance of Taylor's offer. A stock committee was formed, under the chairmanship of Andrew W. Legge, and by October 26, 1928, $20,000 had been subscribed. So in early November, satisfied that this would probably be his best opportunity, Taylor informed the Bradford Board of Commerce that he would accept their offer and move The Taylor Brothers Aircraft Corporation to their city.

In his acceptance letter, Taylor told the Board that, *"About "$17,000 of the stock proceeds will be used to construct a factory building on the airport. It should be a hangar-like structure, 90X100 feet, with a large area of clear floor space for aircraft assembly. This facility will allow us to build at least 100 Chummys the first year, and later one plane per day."* This would prove to be highly optimistic, but in the "boom" business climate of 1928-29, such claims were not unusual. That winter Taylor moved his family into a vacant schoolhouse on the edge of the airport; it would be their home for the next seven years.

In early February, the blueprints for the new Taylor factory had been completed, and bids were being solicited from local contractors. A ground-breaking ceremony was scheduled as soon as the contract was awarded. This activity was going forward even though the stock sale had yet to reach the $30,000 mark.

While Taylor awaited the completion of his new factory, he spent the Spring and Summer of 1929 working on the necessary certification of the B-2 Chummy. If the planes were to be sold they would need an Approved Type Certificate or at the least a "Group II" certification. In May, Taylor spent a week in Washington D. C. going over his stress analysis and construction details with the engineers of the Bureau of Air Commerce.

Following this initial review, the Air Commerce officials directed that a series of flight tests be performed before final approval would be issued. The government test pilots were able to make these flights during the Summer, and the official Group II Approval, Memorandum Number 2-114 was issued on August 27, 1929. This approval covered S/N 9, C492 and two later serial numbers.

In sales literature prepared later in 1929, the Taylor Company offered the Chummy Trainer with the standard Kinner K-5 engine for $3985. A more deluxe model sold for $4375. The plane was also offered with the cheaper two-row Brownback radial engine for $3735 and $4125 respectively. There is no record of any planes being built with the Brownback engines. $4000 was a prohibitively high price for a two-place plane; competitive three and four place models sold for the same or less. It is not surprising that no Chummys were sold.

Adding to the impending financial troubles of the Taylor Company was its ill-considered entry into the Guggenheim Safe Aircraft Competition. In a vain attempt to win the $150,000 prize, X-7303 was taken down from the rafters and given a complicated, (and aerodynamically useless), variable incidence wing. All during the summer and fall of 1929 scarce time and money was diverted to this futile project. Predictably, the plane failed to meet the required performance standards and was disqualified.

At this point in the Taylor Company's history, William T. Piper began to take on a prominent role. Piper was a native of Bradford, his father was one of the oil industry pioneers of the area. Although they endured several periods of economic adversity, his family eventually prospered, holding a number of valuable oil leases. After completing high school in Bradford and serving a short stint in the Spanish American war, Piper was able to attend Harvard University, graduating cum laude with a mechanical engineering degree.

When the Bradford Board of Commerce committee had agreed to sell $50,000 of Taylor Brothers stock, they immediately canvassed their membership for subscriptions. Both Piper and his partner, Lloyd, were "assessed" $400 apiece, and for the rest of his life, Bill Piper would be in the airplane business.

He would soon be more than a minor stockholder in the Taylor organization. In late March of 1929, the company was incorporated in the state of Pennsylvania, and Piper was named as one of the five members of the Board of Directors. Perhaps due to his engineering training, he was asked by the Chamber president, Andrew Burns, to monitor the financial progress of the new enterprise.

It was not long before Piper would need to take an even more active interest. By August, the money raised by the Board had all been expended, the newly completed factory building was not paid for, in fact, the company was insolvent! At this point the Board made a novel arrangement with Piper. They gave him the plot of land the new factory occupied, then in turn Piper took out a loan to pay off the building contractor. As part of the agreement, he would allow Taylor to continue his manufacturing operations in the building.

But by the end of 1929, the manufacturing operations were at a standstill, or nearly so. The final Chummy, NC5E, was finished, but was not sold. The Great Depression was deepening; few people had money to spend in aviation; certainly the prospects of selling a $4000 two place open airplane such as the Chummy were not bright.

The Taylor Guggenheim entry with the variable incidence wing TOM HEITZMAN VIA ED PHILLIPS

The Taylor D-1 glider. Wings look like those on early Cubs Bowers

With nothing else for the plant workers to do, Taylor designed another, quite different, aircraft. In 1988 he told the Author, *"We built a small primary glider, just to give the people something to do, and on weekends we would pull it down the runway, making short flights behind a car. A small crowd would usually gather and we would hear people say "If your glider only had a little engine, it would be just what we would want to buy."*

This glider, identified as Taylor Model D-1, was certainly not the answer to their business problems. Primary gliders were being offered by a number of manufacturers for approximately $400, but few were being sold. It did, perhaps, serve one beneficial purpose. The wings and other structures were built extremely light, perhaps influencing the Cub design that was still to come.

By this time, Piper had become company treasurer, and was using some of his own oil money to keep the factory organization afloat. In fact, on April 30, 1930, he purchased the unsold Chummy C492 from the Taylor Company for $2000, evidently to give a needed cash infusion. But with no further revenues, and none in sight, by midsummer Piper was forced to put the company into voluntary bankruptcy. When the assets were offered for sale, Piper bought them for $761. He reorganized the firm as the Taylor Aircraft Company, and gave C. G. Taylor half-interest, and the title of president. But, significantly, Piper named himself treasurer.

It certainly appeared that C. G. Taylor's efforts to build an aircraft manufacturing company had been a complete failure. He had built only 6 Chummys in three years. One had crashed killing his brother and two remained unsold. Not an impressive record.

With no sales of their expensive Chummy, Taylor and Piper had finally agreed on one thing. To be sold to the cash-starved public, a plane must be cheap, but safe and utilitarian. Their D-1 glider experiment had indicated a need for a very light plane, almost a powered glider, with the ability to carry two people on short, perhaps one-hour flights. To design such a plane was a daunting task, but Taylor knew he must succeed, or his days in aviation were over.

It would be a mistake to infer that the Cub design came "out of the blue", with no knowledge of the competition. Clearly, a number of companies had recognized there was a modest market, even in that Depression economy, for a very light, low cost airplane. Several, including Aeronca, Curtiss Wright and American Eagle, had already entered the low-priced, low-horsepower field.

As he began his layout, Taylor needed to select several items of basic design criteria. First, the plane must be two-place. It was assumed that the typical customer would be a fixed base operator, a large part of whose business would be flight instruction. The plane would at least need to carry an instructor and student. This dictated a gross weight near 800 pounds. As he worked out the design, C. G. came up with these preliminary weight calculations.

Pilot	170 lbs
Passenger	170 "
Gas and Oil	60 "
Engine	100 "
(estimated two-cycle)	
Airframe	300 "
Total	800 lbs

These were a rather remarkable set of figures. They meant that the plane, with an empty weight of 400 pounds, would carry its own weight in passengers and gasoline. It would be quite a departure from the much heavier Chummy design which had a 1500 gross weight limit.

As he mentally formulated the design, Taylor naturally selected some of the familiar specifications used on his Chummys. The USA-35-B airfoil was chosen. It was similar to the Clark Y he had used earlier, giving good lift-drag characteristics at the low speeds expected. For reasons of economy, light weight and stability, a high wing monoplane configuration would be used, the same as had been used for the Chummys, (and later would be used for the Taylorcraft.)

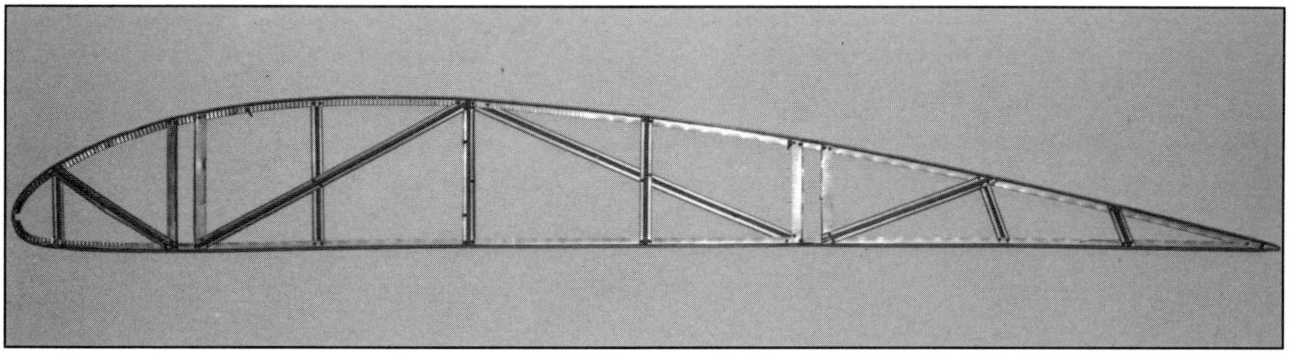

Taylor wing rib AUTHOR

The wing loading would have to be very low, certainly less than 5 lbs per square foot. Using the estimated 800 lbs gross weight as a guide, and a wing span of 36 ft and a chord of 5 ft, the loading came out to a comfortable 800/180=4.44 lbs per square foot. The aspect ratio of 7.20 was well within the desired range. The wing structure, designed next, was a marvel of lightness and strength. Two main spars would be made of spruce, less than one inch thick, the only wood structural members. Compression tubes and tie rods were of steel, of minimum gauge and thickness. The wings ribs were a remarkable new innovation, a departure from the usual built-up spruce and plywood affairs. The members were rolled sections of "Nicral" aluminum, riveted together in a modified Warren truss form. This gave the airfoil a truer shape, it was not subject to warping, and it had a very good strength-weight ratio. A complete rib weighed only 9 ounces!

The ailerons were also of all metal design, extremely light and strong. They were hinged at the top and fastened to a false spar behind the rear main spar. The top aileron horns were made longer than those on the bottom, which gave a differential effect .

In mid July, when Taylor and Piper were discussing the concept of a very light plane, they worked out a cockpit mock-up. Knowing that the horsepower would be extremely limited, Taylor departed from his usual side-by side seating preference, and selected a fore-and-aft arrangement. This reduced the frontal area, and supposedly the fuselage drag. It also greatly simplified the control system. The door to the cockpit, or seating area, was another marvel of simplicity. It was just a long steel framed panel, hinged at the bottom, with two spring loaded pins holding the top. It allowed both pilot and passenger easy entry, and weighed almost nothing.

The rudder, elevators and stabilizers of the tail assembly were again designed with light steel tubing, the outlines being very similar to the Chummy. The landing gear would be a simple arrangement of V-struts braced by diagonal tension struts. The new 7:00X4:00 "balloon" tires had just come on the market; they were light and would reduce the shock of taxiing over rough fields. Byron Shinn was making his cast aluminum aircraft wheels down in Butler, Pennsylvania, so they were selected. A simple shock chord arrangement allowed the diagonal struts to dampen the taxiing and landing stresses further.

The basic design was finished in late July of 1930. Now it had to be translated into a real, flying airplane. Materials, such as tubing, spruce lumber, fabric and dope, were still available, left over when the Chummy production ceased. The only workers left, C. G. Taylor, his father Arthur, and Ken Tibbits, set about to build the world's first Cub.

Wing structure PIPER MUSEUM

C. G. Taylor and Bud Havens with the first Salmson-powered Cub BYRON SCHINN

In a 1988 interview, C. G. described that day. *"It was on a Saturday in late July. I marked the fuselage design in chalk on the cement floor of the shop. My father Arthur cut the tubing and fitted it to the layout, then Ken Tibbits tack welded it together. We welded the two sides together first, then put them on sawhorses and welded in the crossmembers. We finished the main fuselage structure in one day, but it wasn't an eight hour day!"*

Over the next three weeks the various other components were fabricated. Wing spars were sawed out of clear Sitka spruce, then planed to the minimum thickness required, producing what C. G. Taylor always lamented as *"the most expensive sawdust in the world."* The tail surfaces and landing gear were welded up. Finally, the cotton fabric cover, sewn at home on her domestic sewing machine by Mrs. Taylor, was stretched over the frames and given a minimum four coats of nitrate dope. Final assembly took place in early August.

But the choice of a suitable engine still had to be made. The Aeronca Company were producing their little two-cylinder model, but had no interest in selling it to a potential competitor. Ed Heath was flying his single seat Heath Parasols with a four-cylinder converted Henderson motorcycle engine, but just barely.

It does appear that Taylor was planning to use one of the new small engines offered by the Light Manufacturing and Foundry in Pottstown, Pennsylvania. They had already been advertising the availability of the "Brownback Tiger" radial in the Chummy, a larger engine made by the same company. Their "Tiger Kitten 30" offered 25 HP with a weight of only 95 lbs, but being of a geared, two cycle design, it was a radical departure from the typical aircraft engine practice. The name "Cub" also appears to be associated with the use of this engine. According to C. G. Taylor: *"We were standing around one day trying to decide what to call the little plane when Ted Wells said 'Well, if the engine is a Tiger Kitten, then the plane should be a Taylor 'Cub'. That name sounded catchy, so we used it"*

On August 20, 1930, Mr. Gillette, the Taylor Company's Sales Manager, wrote the Department of Commerce requesting an identification mark for *"the first of our new model, the Taylor Cub".* The letter also indicated they expected to use the Tiger Kitten engine. But on September 20, the same Mr. Gillette wrote the Department of Commerce again saying: *"Our August 20 application was based on the installation of the Tiger Kitten Model 30 Engine. This engine was used in test flights here recently, but proved to be of insufficient power, so thus we have had to change the engine."*

"TAYLOR CUB"

MODEL E-2

PERSONAL AND TRAINING PLANE

SPECIFICATIONS

Span	35 Ft. 2 In.
Length	22 Ft. 9 In.
Heighth	6 Ft. 6 In.
Weight Empty	450 Lbs.
Useful Load	400 Lbs.
Gross Weight	850 Lbs.
Capacity	Pilot and One Passenger

PERFORMANCE

High Speed	72 M.P.H.
Landing Speed	26 M.P.H.
Cruising Speed	62 M.P.H.
Cruising Range	240 Miles
Climb at Sea Level	750 Ft. per Minute

POWER PLANT

Continental A-40, 35-40 H.P. at 2500 R.P.M.
Tiger Kitten, 40 H.P., at 1800 R.P.M.
Fuel Capacity: 8 Gals.
Oil Capacity: 3 Quarts.

STANDARD EQUIPMENT

Dual Controls
Wood Propeller
Full Protection Windshield
Wired for Navigation Lights
Altimeter
Tachometer
Oil Pressure
Oil Temperature

EXTRA

Full Cockpit Enclosure
Special Paint Jobs
Other Equipment

PRICES F. A. F.

With Continental Motor	$1,325.00
With Tiger Kitten	1,295.00

Manufactured by

TAYLOR BROS. AIRCRAFT CORP.

BRADFORD, PA.

First Taylor "Post Card" ad. Note the price with a Tiger Kitten engine. PIPER MUSEUM

The first production Cub S/N 12 NC10594 Note the early model Continental A-40 with small fins on the cylinder head. PETER BOWERS

After the failure of the Tiger Kitten, Taylor purchased a French-made Salmson engine for installation in the new plane. It was a little jewel, a nine cylinder radial built like a Swiss watch. But it had serious drawbacks. First, it was too expensive for use in the low price market they were expecting to exploit. Second, as an import, service and repairs would be difficult to obtain. Finally, its metric dimensions and the fact that it *"turned backward",* made it unacceptable in the American market. Taylor recalled that getting a special propeller for this engine presented no problem. *"The Sensenich brothers drove up from Lititz, Pennsylvania, with their tools in their car and carved a prop on the spot. They did this several times for us".*

With the smooth-running little Salmson engine, the Cub proved a delight to fly. At last the Taylor Company had a plane that could be marketed in those depressed economic times. Demonstrations started almost immediately. A number of local pilots took flights and soon several flying clubs were formed. By joining forces in a "Club" several fliers could better afford the modest cost of such a plane. Rensselaer C. ("Bud") Havens, the company test pilot, took the plane forty miles west to Warren, Pennsylvania, and gave demonstration flights to over thirty passengers. As a result of this, E. M. Ferris, of Warren Airways, was signed up as the first Cub dealer, and ordered the first production ship.

Throughout October, Bud Havens was kept busy flying the perky little ship to several East Coast states He claimed to have let over 500 pilots fly the plane, although this seems like an impossible number. At any rate, he garnered orders for nine production planes, and established three dealerships.

Now the Taylor Company had a fine little plane, they had proved it would fly well and could be sold, but they had no proper engine! Taylor recalls their search:

Cub Landing Gear WHITTIER

The first A-40 powered Cub, S/N 12, NC-10594, restored by New Piper Aircraft NEW PIPER

In the early Spring of 1931, the first Continental powered Cub was built, S/N 12, Identification number 10594. A-40 S/N 132 was installed. This plane was sold 4-9-31 to a Mr. Jos. Smith of New York State, before either the plane or engine had received an Approved Type Certificate. Miraculously, this "First Cub" has survived. It is now owned by the New Piper Aircraft Inc. and displayed at their Vero Beach, Florida, headquarters.

The second E-2, S/N 13 is shown as being sold to Ken Tibbits, the factory foreman on 6-11-31. This probably meant it was being kept as a factory demonstrator. The third Cub, S/N 14, was exported to Canada, carrying Canadian registration CF-ARA.

Now that he had a suitable engine, Taylor needed to obtain Department of Commerce approval of the new plane. As he had with the Chummy, he applied

"Piper and I chased all over the country looking for a suitable engine. We heard of the Continental A-40 being built up in Detroit, so we went up there to look it over. It seemed to be just what we wanted so we bought one and brought it home tied to the bumper of Piper's car. R. C. "Bud" Havens was our test pilot at the time we installed the A-40 in our second E-2. Between us, we had 26 forced landings in the first 30 days. It kept blowing gaskets and the single magneto would often quit. However, Continental worked with us to refine the engine and it turned out to be quite reliable."

Finally, Taylor had a plane that could be produced and sold in quantity. The A-40 reportedly cost $400, much less than any other aircraft power plant at the time. It had power enough (but barely) to fly the Cub with a pilot and passenger aboard. A list price of $1325 was quickly published, a sum within the reach of many would-be aviators.

E-2 Cub production at Bradford TAYLOR

for a Group II Approval, receiving #2-358 on 6-15-31. Thus the Cub became the first plane to receive Department of Commerce approval using the A-40 engine.

These "Group II" Cubs were different in several respects from the ATC models that followed a few months later. They carried fuel in two triangular tanks hung in the wing root. The ailerons had diagonal ribs. The windshields had a steeper angle than the later models. The seats were further forward and quite cramped. The engine cowling was in two pieces, top and bottom, with no separate nose bowl.

E-2 Cockpit showing the throttle rod, trim rope and gas shut-off AUTHOR

But other than the minor modifications made (S/N 26 and after), when ATC 455 was attained, the Cub structure remained almost unchanged for the 60 years of its life. The fuselage was made of welded steel tube, mostly 1020 specification, but using 4130 in several highly stressed points, such as below the door on the right lower longeron. Fuselage sides were assembled in a jig which held the right hand assembly on one side, and the left hand on the other. When partially welded, these side assemblies were taken to the main assembly jig where the cross struts were inserted and tack welded in place. The entire fuselage was then removed and put in a rotating fixture for final welding.

The light formers of the turtle-back and front cowling were attached by brazing, using similar welding techniques. The brass brazing rod would run at a much lower temperature than the steel materials, making the process simpler and causing less distortion of these very light-gauge members.

The wing spars were of the plank type made by planing spruce stock down to size, then cutting the end tapers. Drilling jigs were used to accurately locate the bolt holes in the spars.

Next the ribs were slipped onto the spars located by means of a jig table on which the wings were assembled. The compression tubes and the drag wires were then installed and adjusted. In order to keep the wing as light as possible, only thirteen ribs were used in each wing. Very thin, (0.016), aluminum leading edge material was formed and fastened to the ribs with sheet metal screws. The wing had nearly square tips, giving it an antique look.

E-2 Advertising photo TAYLOR

THE TAYLOR CUB

Approved by the Department of Commerce
Powered with a Continental
A-40 Motor

Top Speed
78 m.p.h.

Cruising Speed
65 m.p.h.

Landing Speed
26 m.p.h.

Weight, Empty
525 lbs.

Useful Load
400 lbs.

Gross Weight
925 lbs.

$1,325

ECONOMY — SAFETY — UTILITY

The three big things that every airplane manufacturer has been striving for—All are now found in the TAYLOR CUB.

Its low first cost, small gasoline and oil consumption (25 miles to the gallon) assure unequalled economy.

Its low landing speed (26 M. P. H.), Continental engine and unusual stability make the CUB the safest plane that flys today.

Ability to land in and take off from a small field gives it usefulness far surpassing the ordinary airplane.

The TAYLOR CUB at $1325 F.A.F. Bradford, Pa., is the lowest priced two-placed Government approved plane in America.

SPECIAL FEATURES

STABILIZER ADJUSTABLE FROM COCKPIT
OVERSIZE BALLOON TIRES
FULL PROTECTION WINDSHIELD
METAL RIBS

Dealer Franchises with closed Territory still open to Reliable Parties

---

TAYLOR AIRCRAFT CO.,
Harri Emery Airport, Bradford, Pa.

Gentlemen:

Please send me detailed information regarding your TAYLOR CUB, and your sales plans.
I am interested in owning a CUB ☐
I am interested in your dealers' proposition ☐

NAME..

ADDRESS...

One of the first Cub advertisements AERO DIGEST

The ailerons, of formed aluminum, were assembled in a similar manner. They were huge by lightplane standards, nine feet long and with a chord of thirteen inches. They were hinged at the top, an arrangement that in flight caused some adverse yaw, but with such a slow plane it was acceptable.

In order to increase the revenue per plane slightly, Taylor raised the list price of the plane from $1325 to $1425 in the summer of 1932. This seemed in line with most of the competitors; the Aeronca sold for $1730, the Curtiss Junior for $1595, and the American Eaglet's price was $1575. Both the Eaglet and the Curtiss Junior used the more expensive Szekely engine.

Since marketing had always been his special forte, Bill Piper saw to it that the Cub was well advertised in the aviation press. Costly as they were, several full page ads were placed in national aviation publications. Aimed at the airport operator, they were backed up by personal selling. When Cub production started in 1931, Bud Havens toured the northeastern airports, demonstrating the new plane. Later he went further afield, offering a dealership to almost anyone who would buy, (and pay for), a demonstrator model.

While the flying public enthusiastically accepted the concept of Taylor's little "flying flivver," actual sales grew at an agonizingly slow rate. By the end of 1931, the Cub's first year of production, 22 planes had been manufactured and delivered to customers. 22 more were sold in 1932, but in 1933 sales dropped to 17 units. This hardly indicated a thriving aviation enterprise. With the plane selling for about $1200 wholesale, the total year's revenue amounted to less than $30,000. The direct cost of bought-out items such as the motor, wheels, propeller, fabric and instruments, amounted to over $600 per plane. This left less than $600 per unit to pay for direct labor, rent, advertising, and the salaries of the principals. It is not surprising that Bill Piper continually needed to pay company bills from personal funds. In fact, by the end of 1935, he claimed to have "loaned" the aircraft company over $100,000.

However, to be fair, the sales record of the Taylor Company must be measured against others operating in those depressed economic times. Government records show that in 1934 only 853 civilian light aircraft were sold in the United States. Taylor sold 71 of these, or a bit over 8 %. Their company was one of only five who sold more than 60 units that year.

Sales improved greatly in 1935 when a total of 212 planes were delivered. This increase in sales was due to several factors, most notably the adoption of a "time payment" plan by the Taylor management. Following a marketing strategy long used in the auto industry, the Cub was now offered for $475 down, the balance in 12 monthly payments. This allowed the fixed base operator to use the plane to "make it's own payments", as it would often do.

While Piper was resisting design changes so as to simplify serial production, Taylor was exploring a number of "improvements". Dissatisfied with the poor cruise and top speed performance of the Cub, he designed and built a set of wings which used the more modern 23012 airfoil. When installed on the Cub the cruise speed increased about 10 MPH, but Piper decreed, "We don't need them", so they were hung on the shop wall, later to be burned in the fire. This disagreement over the basic design of future aircraft ultimately would lead to Taylor's departure from the firm.

In the fall of 1935, pleased with the increased sales volume, Piper relented somewhat from his "Don't change anything" edict. He agreed to a modest redesign and upgrading of the E-2 Cub, the first since it was introduced more than 5 years before. The full story of these changes, and the serious problems they caused will be told in Chapter 9, which covers the Taylor J-2.

Then in January of 1936, production of the E-2 Cub would cease. A total of 351 planes had been produced, enough to have made a lasting impression on the flying public. In time, the E-2 Cub would take its rightful place in aviation history as the proud patriarch of that famous aircraft family, "THE CUBS".

Jeannie Hill taking off on skiis in her E-2, SN 60, NC-13179 HILL

Specifications
Taylor E-2 Cub
Gp. II #2-358, 6-15-31 and ATC #455, 11-7-31

PIPER CUB E-2, 2 POLM, 2-358

Engine	Continental A-40 37 hp
Propeller	Wood
Fuel	9 gals.
Oil	3 qts.
No. pass.	1
Baggage	None
Standard weight	925 lbs.
Spec. basis	Aero. Bulletin 7A, Section 3
Serial Nos.	12 to 25 eligible

PIPER CUB E-2, 2 PO-CLM, ATC 455

Engine	Continental A-40, A-40-2, A-40-3 (or A-40-4 on serial Nos. 66, 72 and up having demountable engine mount)
Placard limits	(A-40) (A-40-2) (A-40-3): Maximum, except take-off -- in.Hg., 2550 rpm (37 hp); Take-off (one minute) -- in. Hg., 2550 rpm (37 hp) (A-40-4): Maximum, except take-off -- in.Hg., 2575 rpm (40 hp); Take-off (one minute) -- in.Hg., 2575 rpm (40 hp)
Propeller	Wood
Placard speeds	Not required
Fuel capacity	9 gals. (One tank in fuselage)
Oil capacity	1 gal.
No. passengers	1
Baggage	5 lbs.(Provided approved compartment is installed aft of cockpit - otherwise 0 lbs.)
Weights	Empty use actual (Approx. 556 lbs. with Class I items) Standard 970 lbs.
C.G. limits	(+16) in. and (+20.5)
Spec. basis	Approved Type Certificate No. 455
Serial Nos.	26 and up mfrd. prior to March 24, 1937 eligible. Approval expired as of that date.

EQUIPMENT: (Datum is rear of wing leading edge)(* Means net increase)

Class I.
- 101. 7.00-4 wheels
- 102. Tail skid

Class II.
- None

Class III.
- 301. Cockpit enclosure 7 lbs.
- 302. Steerable tail wheel and 6.00-2 tire (Goodyear-Aircraft Associates installation. Dwg. E-4) 3 lbs.*
- 303. (a) 8.00-4 wheels with 8.00-4 2-ply tires
  (b) 16x7-3 wheels (Goodyear 3LNBM) . No change in weight
- 304. Skis
  (a) Piper S-1000
  (b) Air Transport A or 1220-480
  (c) Federal SA-1
  (d) Marston MFS-1200

NOTE 1. Manufacturer's serial numbers 66, 72 and up have demountable engine mount.

NOTE 2. The wing spars on serial numbers 84 and up have straight butts with "U" shaped reinforcements instead of triangular shaped butts with hole drilled thru spar for fuselage attachment bolts.

NOTE 3. The airplanes eligible are also approved with alternate seats, heel wells, and rudder pedals in accordance with dwgs. J2-A46, J2-A62 and J2-A65. Installation may be made in field using parts furnished by manufacturer.

NOTE 4. Eligible for export to all countries except Great Britain, Canada, and Australia. (May 14, 1938)

Chapter Five

Alexander Flyabout

The Alexander Flyabout. Pictured at Colorado Springs, Colorado. The Approved Type Certificate, #439 was received on 7-31-31 JUPNTER

The A-40 powered "Alexander Flyabout D-1" was the Alexander Film Company's last desperate attempt to hold off economic disaster. By the summer of 1930, the Great Depression had slowed the nation's aircraft production to a near standstill. Alexander could no longer sell their obsolescent biplanes, and their new "Bullet" was having development troubles. Like many of its competitors, the company sought to produce a smaller, cheaper lightplane, more suited to those dismal economic times.

The Alexander Film Co., located in Denver Colorado, was a successful producer of commercial "film shorts" for the advertising industry. J. Don Alexander, their president and an aviation enthusiast, purchased one of the first "New Production" Swallow biplanes in 1924. Finding the plane useful for sales and promotional activities, he soon decided to begin building planes on his own, especially designed for the high altitude conditions in their area.

In the fall of 1925 the company introduced its first model, an OX-5 powered biplane christened "The Eaglerock." Powered by the World War I surplus Curtiss OX-5, it became, in April of 1928, the first plane to receive an Approved Type Certificate with that venerable engine, (ATC # 7). Using the engineering talents of Dan Noonan, soon assisted by Al Mooney, the company became famous for their big workhorse biplanes, which, by using a wealth of wing area, and the lifting qualities of the "Clark Y" airfoil, gave exceptional performance in the higher altitudes of the mountain states.

One design peculiarity gave rise to the common Eaglerock nickname. Contrary to popular design arrangements for biplanes, the Eaglerock had no center section. Thus, when equal length wing panels were used, the lower wing span was about two feet longer than the upper, suggesting the designation, "long wing Eaglerocks".

Eaglerock OX-5 powered "Long-Wing" biplane, 1927 JUPTNER
Note the "spreader bar" landing gear

Those who didn't like this odd-appearance could order a "Combo-Wing". On this model, the lower wings were shortened to give them the same span as the uppers.

The biplane design put forward by Noonan and Mooney was conventional in most other respects. Two passengers were carried in a roomy front cockpit, a pilot in the rear. The fuselage was constructed of welded steel tubing; the wings had spruce spars and spruce built-up ribs. Ailerons were installed in both top and bottom wing panels, giving excellent lateral control. On the first models, a Jenny-type spreader-bar landing gear was used, but later models sported a more modern split axle type. The long-wing planes had 360 square feet of wing surface and weighed 1460 lbs empty. Compared to other planes being built in 1927, the Alexander product was heavier and had more wing area. This extra wing area explained the excellent performance record of this model in higher altitudes. The price at the factory was set at $2475, later raised to $2750. This was competitive to Waco, American Eagle and other OX-5 powered biplanes of the time.

Using the marketing skills that had served the company so well in it's film business, the Alexander company began an extensive sales campaign, starting in 1927. Factory-sponsored entries were made in the Class B division of the National Air Derby, New York to Spokane. Pilots Les Miller and James Charles came in 2nd and 3d in this division, a creditable showing. Cloyd Clevinger was sent out to demonstrate a Whirlwind-powered long-wing; its outstanding performance generated considerable interest. (Clevenger later became the company test pilot, also assisting in the development of new products.)

Under the direction of James McInaney, a sales force was formed and advertising placed in the various trade magazines. As a result, 207 planes were built in 1927, and the crowded Denver manufacturing facilities had to be expanded.

This Model A-2 was introduced in 1928 JUPTNER

This success prompted Don Alexander to further increase the company's investment in the aviation field. In 1928, a new plant was constructed in Colorado Springs, about seventy miles south of Denver. When completed in the summer of that year, it was said to be one of the largest aircraft manufacturing plants in the country. The company reported that it had the capacity to build one plane per hour. (Assuming an 8-hour day, this would amount to about 2000 units per year!) New models were also introduced. Al Mooney, now chief engineer, cleaned up the design of the earlier biplane models, and introduced the A-1, A-2 and A-3. The chief difference was the installation of a wide, (6-foot), center section, which gave the plane a much more conventional appearance. Better streamlining improved performance.

This new A model was offered with three engine options. The A-1 used a Wright Whirlwind J-5 of 220 HP, making it a very lively

The Alexander "Bullet", a plane far ahead of its time. Development troubles and the onset of the Great Depression halted production. JUPTNER

performer. The $7500 list price reflected the higher price of the Whirlwind motor. The A-2 model had the familiar Curtiss OX-5 installed, the price set at $2475. A similar model, the Hispano-Suiza powered A-3 sold for $3250; later this was raised to $4000. A wide array of other engines were offered; the Comet, Curtiss Challenger, Warner, etc., with model numbers running up to A-13.

Reflecting the fact that the World War I surplus engines were fast disappearing, the Eaglerock Model A was also offered without engine for $2250, which included cowling and installation for either your OX-5 or Hisso engine. This price list points out the dilemma facing all aircraft manufacturers at this time. The supply of cheap engines was drying up, and there were no really suitable replacements. This would ultimately speed the search for a small, inexpensive aircraft engine, and the introduction of the Continental A-40.

However, none of these economic forces were apparent to the Alexander management during 1928, and the early months of 1929. Sales continued to increase, perhaps as many as 500 planes being built in 1928, and a sizeable number in 1929. Flushed with optimism, and perhaps sensing that the 3-place open biplane was becoming obsolete, they directed Al Mooney to design a radically different plane, a 4-place cabin model called the "Bullet" with a cantilevered low wing, and retractable landing gear. It was said to be the first low wing cabin monoplane to receive an ATC.

At the time Mooney was designing the Bullet, the cantilever wing concept was just becoming popular, perhaps due to the success of the various Fokker monoplanes. Boeing was building their Monomail; Lockheed had introduced the Sirius and the famous Vega. Even in remote Sioux City, Iowa, Swen Swanson had designed the little Kari-Keen with a cantilever wing. The Bullet wing was made in three sections, a center section of constant chord, and two outer sections of tapered plan form. The tail section was of welded steel tubing, the entire plane covered with cotton fabric.

The Alexander executives introduced the new plane to the public during the 1929 Detroit Air Show. It generated substantial interest from the aviation community, but unfortunately, few sales resulted. When making flight tests for the ATC approval, the plane was found to have some unpleasant spin characteristics. Al Mooney's engineering crew seemed unable to remedy the problem, so management called in Ludwig Muther, a former Fokker associate, to help. At this point, Mooney left, eventually going on to design the Culver Dart, and ultimately his own line of sleek low wing monoplanes.

After consulting with the famous aero engineer, Dr. Max Munk, the spin problems were finally solved, and the Bullet received its Approved Type Certificate on 5-6-30. But the fix had involved lengthening the fuselage by almost five feet, with a resultant increase in both weight and cost. When finally offered to the public in the summer of 1930, the gross weight had gone up to 2790 lbs, and the list price was a lofty $7200. This was more than the depression market could stand; only one C-7 model was produced, NC309V, S/N2013.

Clearly, to remain in business, the Alexander company needed to offer a less costly plane to the aviation public. This led to the Flyabout design, one of the first planes to use the new Continental A-40.

Dr. Ludwig Muther left the company in mid-1930 when the Bullet sales failed to materialize. A young junior engineer, Proctor Nichols, was named Chief Engineer and given the responsibility for the design of this new lightplane. The result, when introduced at the Detroit Air Show in 1931, was a rather conventional high wing monoplane. In order to gain as much lift as possible, the wing was made almost 38 feet long. With a 5 foot chord this gave 175 sq. ft. of lifting area. The airfoil chosen was the Munk M-6, suggesting Max Munk may have had some connection with the project. A side-by-side seating arrangement was chosen, but with the fuselage only 30 inches wide; it must have been a bit crowded. The door was not covered on the inside in order to gain perhaps an inch more space, and the two seats were staggered four inches to make the best of this situation.

The cabin was completely enclosed with large pyralin windows, including a skylight in the roof. A wide door was installed on each side. Dual stick and rudder controls were furnished. The fuselage was finished with six coats of green pigmented dope, the wings in cream.

The split-axle landing gear used Goodyear 16X7-3 airwheels which were also the only shock absorbers. The empty weight came to 573 lbs, and with a gross allowance of 962 lbs., the payload became 389 lbs. Only 7 gallons of fuel could be carried in the fuselage tank, allowing a range of 175 miles. Clearly, this was not intended to be a cross-country airplane. The cruising speed was advertised as 70 mph, the top speed 80 mph. Dual controls and a cabin heater were offered as standard equipment.

It seems likely that the design was originally conceived with the Continental A-40 in mind, because it was the engine installed when the prototype was completed in 1931. It became the first plane fully certificated with the A-40 when its ATC was granted 7-31-31. (The Taylor Cub had received its Group 2 approval a month earlier on 6-11-31.)

But when the first demonstrations of the new plane were held, its performance in the 5000 ft. altitude of Denver doomed it to failure. With one aboard it would get off the ground and climb slowly, but with two aboard, even getting in the air was questionable. Nichols, the designer, boasted of flying the plane over Pike's Peak at 15,000 feet, which may have been true, but he certainly had no passenger.

This Flyabout, with Canadian registration letters, is a later model with a Szekely motor. JUPTNER

THE ALEXANDER FLYABOUT

FROM the standpoint of sale-ability, that article which most closely conforms to the ideals of the public, is most likely to succeed. It was with this in mind that a careful survey of present market conditions and design trends was made before any actual engineering work was started on the new project.

The present trend toward lighter, lower powered planes is based on a definite public demand and economic necessity. It was only by the introduction of the small automobile at a price within reach of the general populace that motoring was made popular. It is logical to assume, therefore, that the light, highly efficient airplane is here to stay.

The Flyabout was designed to fulfill the demand which exists, for a plane combining low first cost, economical operation, good performance and comfort. As the job is intended primarily for student training work and for solo students, the cabin features are most desirable, in that the largest percentage of students will be flying cabin equipment on the completion of their training.

Visibility was considered of prime importance and windows are so arranged that when opened the cabin becomes an open cockpit. Dual controls are provided, either set being removable. The seating arrangement, being side by side, with a slight stagger, provides the utmost convenience for instruction and is the most companionable arrangement for cross country flying. The four-inch stagger increased the comfort of the already spacious cabin.

THE Flyabout is flown on 175 square feet of wing area, the wings being attached to the top longerons by four fittings at the corners of the cabin. Faired lift struts of rounded seamless steel tubing form the external bracing. The front spars are of wood, and are built up box beams, with spruce cap strips and ¼-inch spruce planking. Rear spars are solid rectangular spruce beams made up of two laminations of grade A spruce glued together. The internal bracing consists of twelve single drag bays, six in each panel. At strut connection points metal drag struts are used. Drag struts at other stations are of spruce.

SPECIFICATIONS

Wing span	37′ 9″
Length Overall	21′ 4⅝″
Height Overall	5′ 8½″
Chord	5′
Incidence	2½ degrees
Dihedral	½ degree
Gasoline Capacity	8 gals.
Oil Capacity	1 gal.

AREA

Wing	175 sq. ft.
Ailerons	20 sq. ft.
Stabilizer	15.1 sq. ft.
Fin	3.1 sq. ft.
Elevator	8.12 sq. ft.
Rudder	7.5 sq. ft.

WEIGHTS

Total Weight Empty	536.5 lbs.
Crew	340 "
Fuel and Oil	55.5 "
Baggage	30 "
Useful Load	425.5 "
Gross Weight Loaded	962. "

LOADINGS

Wings	5.497 lbs.
Power	27.48 "
Load Factor	6.5 "

PERFORMANCE

High Speed	85 m.p.h.
Stalling Speed	30 m.p.h.
Landing Speed	34 m.p.h.
Take-off (distance)	175 ft.
Rate of Climb	650 f.p.m.
Cruising Speed	75 m.p.h.
Cruising Radius	175 miles
Fuel Consumption—g.p.h.	3 gals.
Fuel Mileage (cruising)	25 miles per gal.

STRUCTURE

Fuselage, welded steel seamless tubing. Fabric covered.
Landing Gear, Air Wheels. Split Type.
Wing, M-6 Airfoil section, box spruce spars, spruce ribs, fabric covered.
Side by side arrangement, dual control, two doors, pyralin windows, allowing open and closed effect.

EQUIPMENT

Air wheels, switch, tachometer, altimeter, oil gauge, fahrenheit, first aid kit, Pyrene fire extinguisher, center throttle control and tie down ropes.

POWER PLANT

Continental A-40, four cylinder opposed, 35 horse-power motor at 2500 r.p.m. Scintilla Magneto. Zenith Carburetor.

Spruce truss type ribs, are used, the space being 18 in. Three false nose ribs are used between each rib, to insure the trueness of the air foil on the leading edge.

The fuselage is constructed of aircraft steel tubing, chrome molybdenum being used wherever it serves to reduce weight.

Two very wide doors are provided giving access to the cabin on either side. The doors are made in the form of a hollow section in order that still greater width may be secured at the seats.

Full air wheels are provided, and are used to absorb landing and taxiing shocks. Original experiments were conducted with rubber cord shock struts in addition to the air wheels, but it was felt that these were unnecessary. The chassis is of the conventional split type with a four-foot six-inch tread.

The Continental A-40, 38 H.P. motor is used for the power plant, and is installed in the nose, with a tractor type propeller. With front windows open the pilot may crank his own motor, from the rear, without removing his hand from the engine controls on the dash, and with no danger to himself. An eight and a half gallon gasoline supply give a two hundred and forty mile cruising radius. The gas tank is located immediately behind the firewall and under the instrument board.

The fuselage is beautifully finished with six coats of pigmented dope, standard color being green with cream wings. The interior of the fuselage is attractively finished and upholstered.

Tail surfaces are of aircraft steel tubing with channel steel ribs, covered with fabric. Control surfaces are large and provide excellent maneuverability and perfect control in normal flight or stall.

Flying qualities of the ship have been carefully determined and checked by extensive field flying and long cross country trips. A careful check of gasoline and oil consumption was kept and it was found that an average of 22¼ miles to the gallon of gas is obtained.

THE extreme economy of this ship is shown by the fact that the ship has been operated at a total cost of 83 cents an hour for gas and oil. A careful estimate shows that the total cost of operation, including gas, oil, depreciation of plane and motor, mechanical repairs and upkeep, will be less than $1.35 per hour. This totals less than $.02 per mile or less than a cent per passenger mile.

This article, announcing the Flyabout, was written by Proctor Nichols the project engineer.

The little A-40 needed to run wide open to maintain flight at all. It is not surprising that no customers could be found for such a plane. The more powerful Szekely SR-3-0 engine was quickly substituted, and received ATC #449 on 9-5-31. It was a Szekely powered plane that made the 3500 mile round trip to the National Air Race and show in Detroit.

Glenn Miller of Des Moines, Iowa, owned and flew one of these Szekely Flyabouts, NC11085, back in the 1930's. He worked for Woltz Studios, a large photography company that specialized in high school senior class yearbook photos. If they won a contract from a school they would give a free aerial view of the town to the class. (Woltz Studios took the high school graduation photos of the Author's 1937 Paullina, Iowa, class, all thirteen of them.)

Miller used the plane to build up flying time and to occasionally take one of these aerial views. After about 100 hours of use, the Flyabout and its troublesome Szekely engine was sold.

Records indicate that four of the Alexander Flyabout D-1's were manufactured, including the prototype, X-447V, S/N #3000. All were later modified by installation of the more powerful Szekely 45 HP SR-3-O engine, thus becoming D-2 models. The business was closed down in 1932 and sold to a consortium called Aircraft Mechanics, but there was no further Flyabout production.

Specifications
Alexander Flyabout D-1 & D-2
ATC-439 7-31-31, ATC-449 9-5-31

MODELS: Flyabout D-2, 2 PCLM

T.C. NUMBER: ATC 449

Engine	Szekely SR-3 45 hp
Propeller	Wood
Fuel	7 gallons (one in fuselage)
Oil	1 gallon
No. passengers	1
Baggage	3 lbs.
Standard weight	982 lbs.
Spec. basis	Approved Type Certificate No. 449
Serial Nos.	3002 and up manufactured prior to 9-30-39 eligible. Approval expired as of that date.

Class I equipment:
Airwheels

Chapter Six

Heath "Parasol" LNA-40

This 1932 photo shows one of the first A-40 powered Heath Parasols. Note the landing light. JUPTNER

Ed Heath could rightly be called the father of the homebuilt aircraft industry. In the 1920's he built, then marketed in kit form, a small single seat monoplane using a motorcycle engine for power. Now, for the first time, an aspiring young aviator could purchase a kit, or part of one, and at least dream of flying his own plane.

Heath was a native of New York State, born there in 1888. By the time he was in high school, Curtiss and other aviation pioneers were making record flights in his area, so it was natural that he become interested in aeronautics. He is reported to have built and flown his own plane in 1909. About 1912 he moved to Chicago and opened a complete aircraft supply business. The Heath Airplane Company advertised a variety of aircraft products such as fabric, propellers, spruce wood, engines and instruments. He would even build a ready-to-fly airplane to your order.

After World War I Heath was able to add various items of aviation surplus to his inventory. His typical catalogue offered OX-5, Hall- Scott and Le Rhone engines, as well as parts for Standards, DeHaviland DH-4's, Curtiss Jennys and Thomas Morse Scouts. According to one story, it was the Thomas Morse parts that started him in the Heath Parasol business.

It seems that Heath had a number of Thomas Morse wings, and no use for them. He decided to build a simple single place plane using these wings and powered it with a Henderson motorcycle engine. With the 110 pound Heath in the cockpit, the plane barely made it off the ground, but it did fly. And it was cheap to buy and simple to build!

He further refined the design, building a 25' wing with a Clark Y airfoil, which greatly improved the flying characteristics. To make construction of the fuselage simple, he used a "Bolted Body" concept. No welding was required, instead the fuselage tubing was fastened together with rivets, actually shingle nails! Small turnbuckles and aircraft hard wire took care of the bracing. The result was a surprisingly light and strong structure. Because it was a high wing monoplane, he chose the name "Heath Parasol".

Not only was the plane simple to build, it was simple to buy. As the price list on the following page shows, the amateur aircraft builder could purchase parts in separate "kits"; for instance the fuselage kit cost $40.56. Thus a person could get started on the plane and purchase additional parts as his finances allowed. If the kit for a complete plane was purchased, it came in a large cardboard tube.

The "Parasol" sold well throughout the last half of the 1920's, especially after the "Lindbergh Boom". There is no way to determine the exact numbers, since most customers bought only part of a kit at a time. When the planes were registered with the Department of Commerce, and many were not, they might carry the Heath name or the name of the builder. Certainly only a small percentage of the planes begun were ever completed. An even smaller number actually flew.

The Author had personal knowledge of just one Heath Parasol that was actually built and flown. Harvey Christianson, an old time pilot and instructor from Spencer Iowa, built a Heath in 1928. Dean Gilmore converted a Henderson motor for the project. It barely flew, Christianson was rumored to have taught himself to fly in the plane, without previous instruction. The plane was so underpowered, it could only fly on very cool days. Christianson reportedly flew only in a bathing suit, to keep the weight down!

Like most of the aviation entrepreneurs of that era, Heath craved publicity. One of the ways to make news was to enter the various air races which were then popular. Again capitalizing on his small stature, Heath designed several miniature racing aircraft, the most famous of which was the Baby bullet. Powered by a Bristol Cherub engine, it created quite a sensation, usually winning races in its class.

Each year the National Air Races offered an opportunity for sportplane builders to show their wares to the aviation public. So, starting in 1926, Heath entered his little monoplanes, and they usually did well in their field of competition. Ed Heath was personally involved in the design of these various racing models, and usually did the test flying himself.

A stock Heath Parasol NC10373 rounding an air race pylon JUPTNER

The company was already offering a high wing parasol model, and a companion center wing version. In the winter of 1930-31, a low wing model had been designed, and as usual, Ed Heath was making the test flights.

Duke Mueller and his Heath SNA-40 racer, 1931 JUPTNER

HEATH AIRCRAFT CORPORATION

Formerly HEATH AIRPLANE COMPANY, Inc.

LINCOLN 6196-6197 1721-29 Sedgwick Street Chicago, Illinois

Price of Material Needed in the Construction of the "Parasol"

Fuselage	$40.56	Instrument and cowl support	$ 5.12
Controls	8.53	Tail unit	13.59
Pilot seat	2.32	Landing gear	40.76
Tail skid	.77	Center wing supports	3.59
Cowling	4.94	Outer wing supports	10.45
Engine mounting	3.44	Wings, gas tank, ailerons	76.00

TOTAL COST IF PURCHASED SEPARATELY (varies slightly with market) $225.53

SPECIAL PRICE ON COMPLETE BILL OF MATERIAL, $199.00; BOXING, $5.00

Blueprints

Complete set of blueprints, showing construction details and including an itemized list of parts needed, bolted type	$5.00
Welded Fuselage Prints, extra	2.00
Prints of Wings, only	2.00
Prints—Anzani "Parasol"	5.50
Prints—Henderson Conversion	.65
Prints—Pontoons	2.00
Prints—Skis	.75

Notice

Any parts that are found defective may be returned to us upon receiving shipping instructions from us.

In case of damage in transit, notify your express or freight agent at once and have him verify damage.

ALL PRICES SUBJECT TO CHANGE WITHOUT NOTICE

General Specifications of the "Parasol"

Span	25 ft.	Stabilizer area	5.5 sq. ft.	Useful load	300 lbs.
Chord	4 ft. 6 in.	Rudder area	3.8 sq. ft.	Gas capacity	5 gals.
Angle of incidence	1 degree	Length over all	16 ft. 9 in.	Oil capacity	6 qts.
Wing area	110 sq. ft.	Height over all	5 ft. 10 in.	High speed	85 m.p.h.
Aileron area	10 sq. ft.	Weight, empty	285 lbs.	Landing speed	28 m.p.h.
Elevator area	5.2 sq. ft.	Rate of climb (first minute)	600 ft.	Cruising radius	200 miles

Performance of Anzani model practically same as above

Price of "Parasol"

COMPLETELY ASSEMBLED, READY TO FLY AWAY AT CHICAGO

Equipped with **Heath-Henderson** motor	$975.00
"Parasol," complete, but without motor	690.00
Anzani "Parasol," without motor or mount	750.00

TERMS

If crating is required for shipment by rail, a small extra charge will be made, viz.:

Crating complete plane	$18.00
Boxing, motor only	3.00

ALL ORDERS MUST BE ACCOMPANIED BY AT LEAST ONE-THIRD IN CASH; BALANCE C. O. D. OR SIGHT DRAFT.

Export orders must be accompanied by Cash in Full, and a reasonable amount for crating and shipping charges must be included. Personal checks must be certified.

The duty on aircraft material to most foreign countries, such as Canada, is in the neighborhood of 33 per cent.

Price, Complete Bill of Material, $199.00

Heath price list, 1929

One day in February 1931, while flying this new low wing model, a structural failure occurred and Heath fell to his death. With the company founder and chief executive officer gone, the company was reorganized; the name changing from the Heath Airplane Company to the Heath Aircraft Corp. Fred Seiler, who had been Chief Engineer at Kreider-Reisner, was brought in to manage the enterprise. He evidently saw the need for Approved Type Certificates on the Heath planes. For this he hired Charles Morris to do the stress analysis and redesign. Unfortunately, in July of 1931, Fred Seiler was killed while testing one of the experimental models. Morris took over as chief engineer and completed the ATC work. He even did most of the test flying involved. (Brave man). By updating the standard Parasol into the more sophisticated, (and better looking), Heath LNB-4; ATC # 456 for a Henderson-powered model was awarded on December 16, 1931. By coincidence, the Taylor Cub received the preceding ATC, #455.

But the updating of the original Parasol had also added considerable weight to the plane, almost 100 lbs. The little Heath B-4 motorcycle engine could barely get the plane off the ground. More horsepower was needed, and the only option available at the time was the Continental A-40.

The Heath company had already used the A-40 on several of its racing planes. Duke Mueller flew an experimental LNA-40 in the low-horsepower free-for-all and won with an average speed of 91 MPH. When the LNB-4 performance proved to be inadequate, the Continental A-40 was an obvious choice as a replacement.

A restored, (but somewhat modified), LNA-40 at a midwest fly-in. AUTHOR

Like most of its ancestors, the Heath LNA-40 was a single seat open-cockpit high wing monoplane. The fuselage had been generously faired to a more rounded shape, although the cockpit still remained only 18 inches wide. The framework was constructed of welded light gauge steel.

Willis Kysor in an LNB-4, S/N 151, NC10740 JUPTNER

The wing used spruce spars and built up wood ribs. Full length ailerons were added and the span increased to 31 ft. 3 in. A "high-altitude" wing option increased this span to 36 ft. A clipped-wing version was offered for racing. Tail surfaces were welded light steel tubing. The split-axle landing gear used disc type 16 X 4 wheels.

The Heath LNA-40 received its Approved Type Certificate #487 on 6-3-32. When first introduced, the list price was set at $1224; later lowered to $1085 in 1933. A complete kit of factory-made assemblies, less the engine and propeller, could be bought for $499. The A-40 engine was offered separately for $450.

Ed Garber Jr. of Fayetteville NC owns and flies a restored Heath LNA-40, S/N 1006, NC 15792. He tells this history of his plane:

"My Heath was built at the factory as an LNB-4 in 1932, but there is no record of a Henderson engine ever being installed. It was sold the same year for $200 plus $16 for crating. The owner installed his own Continental A-40 engine, making it an LNA-40. However, ATC #487 was lost, so it was licensed under ATC #456 as an LNB-4, then had a Supplemental Type Certificate issued for the A-40 installation. It was owned by Andy Anderson for several years, then a museum in Santee, SC, then returned to Andy. I acquired it from Andy and after a few years in the warehouse, I restored it. This is my third Heath Parasol restoration. It flies quite well if the weather is not too hot."

Ed Garber's restored LNA-40 GARBER

For many reasons, the Heath LNA-40 did not sell. First, of course, was the depressed economy of the 1930's. Also, a single place plane for $1225 was not a good economic choice when a two-place Aeronca or Cub could be had for $300-$500 more. The two-passenger planes could be used for instruction and a source of revenue, a single place could not. Records indicate that a total of nine LNA-40's were built, including two which were experimental.

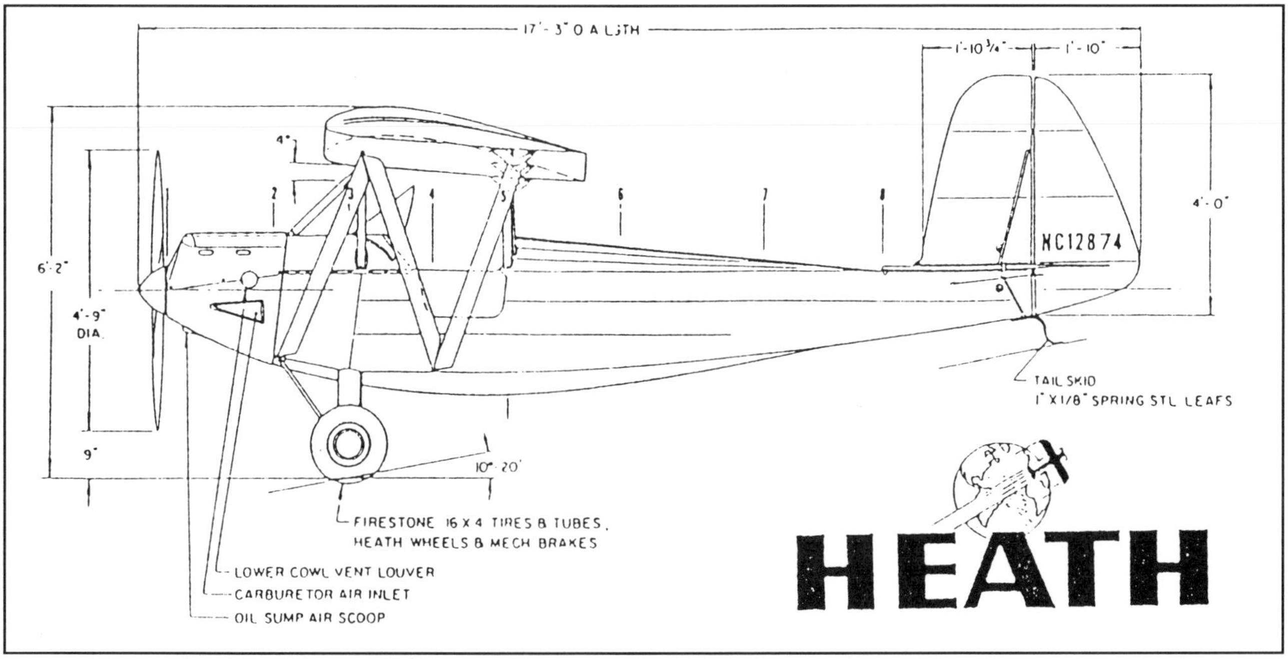

Heath LNA-40 drawing from a factory brochure. HEATH

Specifications
Heath LNA-40
ATC 487 6-3-32

INTERNATIONAL-HEATH LNA-40, 1 POLM, ATC 487

Engine	Continental A-40 37 hp
Propeller	Wood
Fuel	9 gals.
Oil	1 gal.
No. pass.	None
Baggage	4 lbs.
Standard weight	700 lbs.
Spec. basis	Approved Type Certificate No. 487
Serial No's.	160 and up mfrd. prior to 12-1-35 eligible. Approval expired as of that date due to sale to E. H. Anthony (Heath Aviation Company), Benton, Harbor, Michigan

Class III equipment: (* Means net increase) 7.00-4 wheels and tires 2 lbs.*; Brakes 4 lbs.*

NOTE 1. Eligible with or without aileron gap fairing strip.

Heath LNB

(For Comparison)

INTERNATIONAL-HEATH LNB-4, 1 POLM, ATC 456

Engine	Heath B-4 25 hp
Fuel	9 gals. (Two tanks - one 4½ gal. tank in each wing)(See NOTE 2)
Oil	1½ gals.
No. pass.	None
Baggage	16 lbs.
Standard weight	700 lbs.
Spec. basis	Approved Type Certificate No. 456
Serial Nos,	150, 154, 156 and up mfrd. prior to 12-1-35 eligible. Approval expired as of that date due to sale to E. H. Anthony (Heath Aviation Company), Benton Harbor, Mich. (See NOTES 1 and 2)

EQUIPMENT: (* Meant net increase)

Class I: Wood propeller.

Class III: Wheel brakes 4 lbs.; Hand starter 4 lbs.; Propeller - fixed metal 2 lbs.*; 7.00-4 wheels and tires (Shimm) 2 lbs.*

NOTE 1. (a) Serial Nos. 160 and up have revised aileron but do not have aileron gap fairing strip. (These aircraft also have raised wing. Vertical distance from lower front cabane strut bolt to bottom of front spar is 24 inches.)

(b) Serial Nos. 150, 154, 156, 157, 158 and 159 have aileron gap fairing strip installed and the following:

Fuel	5 gals.
Baggage	None
Standard weight	675 lbs.

Equipment: Brakes 4 lbs.; Hand starter 5½ lbs.; fixed metal propeller and 7½ lbs. other equipment not affecting flight characteristics.

Chapter Seven

Heath Center-Wing CNA-40

This Heath Center-wing has just landed on Michigan Boulevard in Chicago. It was to be exhibited at the 1933 Worlds Fair, "A Century of Progress". JUMPTER

The Heath "Center-Wing" CNA-40 was a logical outgrowth of the Heath racing plane designs. In 1928, Ed Heath designed the fast little "Baby Bullet", and powered it with a British-built Bristol Cherub engine of 32 HP. The imported engine was expensive, ($1270), but the investment proved a wise one. When the little plane was first raced at Los Angeles in September of 1928, the top speed proved to be over 140 MPH. Even when carrying a 75 lb weight handicap, the Bullet won the main light plane race, garnering $1500 in prize money. Over the next three years, the winnings totalled over $5000.

After the success of this center wing racer, some of Heath's customers ordered similar clipped wing models for sport use. There may have been further pressure from a competitor, the Church Mid-Wing. In 1928, a Chicago airplane entrepreneur named Church, in what possibly was a violation of copyright law, made copies of the Heath plans, exactly traced copies, but modified the parasol to a center wing. Hundreds of sets of these plans were sold, and at least some planes completed and flown.

The Author at one time owned a homebuilt Church, bought from a mechanic in Rapid City, SD. For a while a restoration was considered, but the poor quality of workmanship dictated the plane be retired safely to a museum.

Gene Chase, of Oshkosh, WI, restored and flew a Church Mid-wing in the 1960's. It was an interesting sight at Midwest fly-ins, the little Henderson engine buzzing along at 3000 RPM. But Gene's flights came to an end one day at El Reno, OK, when the crankshaft extension on the motor broke, and the propeller separated from the airplane. Fortunately, a successful forced landing was made on the airport runway. The plane now resides in the EAA Museum.

When Seiler, and later Morris, readied the center wing design for certification, the racing performance was certainly in mind. Whereas the parasol model had 135.5 sq. ft. of wing area, the CNA-40 ha only 107.4 sq. ft, the wings being clipped to a span of 26 ft. 10 in. The really daring customer could order even shorter wings with only 69.6 sq. ft. of area, obviously only useful for competition. In this configuration the plane cruised at 90 and landed at 60.

The fuselage was of exactly the same construction as the LNA-40 parasol, welded steel tube, with wooden fairings. The A-40 engine installation was also identical. The wings were of standard Heath construction, wood ribs and spars, but shortened as mentioned above. Wing bracing was provided by a sturdy set of steel N-struts, but in this case the angle was quite flat, putting considerably more load on the fuselage and wing fittings. It is possible that this had something to do with the test flight failures noted earlier.

The Center-Wing was offered to the public for $1095 in 1933, when the first models were built. By that time the plant had been moved to Niles, MI, and had been reorganized as the International Aircraft Corporation. In order to capitalize on the 1933 Chicago World's Fair publicity, the governor of Michigan sent three of these Center-Wings to the Fair on a goodwill mission, a greeting from the people of Michigan. The planes landed in the street in front of the main entrance, and remained on exhibit there for some time.

The Author's Church Mid-Wing AUTHOR

Heath "Baby Bullet" 1930 FLYING MANUAL

Sales, however continued to be slow. Two of the Worlds Fair planes eventually found owners, the third went back to the factory as a demonstrator. Two or three more of this model may have been manufactured, but their disposition is not known.

The Heath Company, without any real sales volume, continued to have financial problems. Questionable stock sales and lawsuits resulted in seizure by the Treasury Department in 1935. The assets were sold to Howard Anthony, of Benton Harbor, MI, who eventually changed the entire focus of the enterprise. Instead of building planes, Anthony sold electronic kits; the famous "Heath Kits" used by millions of radio hobbyists.

One CNA-40 model survived and has been restored by the late Bob Thompson of Dayton, OH. Photos of the plane at the 1933 air races, and later as restored in 1975, are shown below.

The Author props Gene Chase's Church Mid-Wing for its last flight, (1975) AUTHOR

NC12882 at the 1933 Air Races. JUPTNER

Thompson's restoration of NC12882, (now 12881), in 1975 THOMPSON

Specificatons
Heath CNA-40
ATC 496 2-2-22

INTERNATIONAL-HEATH CNA-40, 1 POLM, ATC 496

Engine	Continental A-40 37 hp
Propeller	Wood
Fuel	7½ gals.
Oil	3/4 gal.
No. pass.	None
Baggage	None
Standard weight	675 lbs.
Spec. basis	Approved Type Certificate No. 496
Serial Nos,	C-50 and up mfrd. prior to 12-1-35 eligible. Approval expired as of that date due to sale to E.H. Anthony (Heath Aviation Company), Benton Harbor, Michigan.

Class III equipment: 7.00-4 wheels (Heath or Shimm), no change in weights.

Chapter Eight

Rose Parrakeet

The first certified Rose Parrakeet. This is a Model A-1, X13677, S/N 101. Date, 2-18-36 JUPTNER

The Continental A-40 was to power more than a dozen aircraft designs, but it reached its peak in performance when mated with the little Rose Parrakeet biplane. While other models were struggling to lift their two passengers off the ground, the perky Parrakeet was leaping into the sky, climbing at almost 800 feet per minute, and when running flat out, indicating over 100 MPH! It would perform most aerobatic maneuvers with ease and precision, yet was docile enough to be flown by a novice pilot. Handsome as well, its pleasing biplane layout became a classic among model builders.

Jennings Wilson "Jack" Rose, the man responsible for this tiny aeronautical gem, was born in the Chicago area about the turn of the century. His first aviation experience came in June, 1923, when he enlisted in the U. S. Army as a flying cadet. (Charles Lindbergh was in the class following.) He was sent to Brooks Field, near San Antonio, where he began the pilot training course. However, after only 4 months, and 13 hours of flying, including 1 hour of solo time, he was discharged. Undoubtedly disappointed, he returned home to his machinist profession, but also continued to fly. His log book shows he soloed, again, in a Curtiss Oriole on 8-28-24. From 1924 through 1929 he evidently "barnstormed" in his spare time; his log book shows many short 20 minute flights in such planes as JN4-D, Canuck and a Hisso-powered Lincoln Standard LS-5.

Keeping these old surplus WWI planes flying certainly required considerable mechanical ability. Reflecting this experience, Jack received his Mechanics ID in 1928. He went on to receive his Transport License #9687 in 1929. He finished up the "Roaring Twenties" flying a new production Waco 10.

About this time he began to think of designing and building his own plane, a small single place sport biplane. There are a number of theories as to the parentage of the Rose Parrakeet. Some say it is a near copy of Swen Swanson's SS-3 Lincoln Sport, plans of which were published in the 1920's. Certainly the plan form, the overall dimensions and the underslung lower wing confirm this. But the rudder is similar to a Fairchild 22, and the landing gear was actually taken off a Buhl Pup. In the air, it had the looks of a baby Waco.

Jack himself said: *"It contained at least 10 good features from other airplanes."* Whatever its ancestry, it turned out to be a beautiful little machine, an esthetic triumph if not a financial one. It was powered by a Heath-Henderson motorcycle engine. Both being in the Chicago area, Heath and Rose had become friends, so it is not surprising that the first powerplant tried in the little plane was Heath's B-4. Rose says, *"Heath did not make up the motor, we purchased the conversion parts from him and followed his instructions as best we could."*

The wing span was 20 ft, the length 17 ft 6 inches and the height 5 ft 8 inches. The empty weight was only 410 lbs, about the weight of the early Heath planes. Wing area was 116 sq ft. If the Henderson engine was rated at 27 HP, this gave a power loading of 15.2 lbs/HP. With the gross weight set at 675 lbs, the wing loading was a relatively light 6.15 lb/sq ft. Ten gallons of gasoline was carried in a fuselage tank.

Rose used the common construction practices of the time. The fuselage was a welded steel tube framework, faired with spruce strips, and covered with fabric. The wings had solid spruce spars, ribs of plywood and spruce, and were also fabric covered. It claimed a special Rose-designed airfoil, but it was actually almost an exact copy of the NACA 2412. Interplane and cabane struts were streamlined tubing, braced by streamlined landing and flying wires. Ailerons, on the lower wing only, were controlled by cables within the wing. The tail group was made up of light welded steel tube, covered with fabric and braced with a single set of streamlined wires. The landing gear was a split-axle type, shock cord dampened, and equipped with 16X4 tires, also from Heath. In order to save weight, Rose cast his own 16 inch aluminum wheels.

Standard equipment included a Gardner wood propeller, fire extinguisher, Arens engine controls, ignition switch, primer, aluminum engine cowl and propeller spinner. Only four instruments were furnished; an altimeter, tachometer, oil pressure and oil temperature. (In those days, an airspeed indicator was not required.)

This first model was finished in 1930, but did not receive registration number NX12084 until 1932. Rose made several short flights with the plane, but results were not satisfactory. The engine would continually overheat, causing a number of forced landings. Chapter 1 detailed some of the overheating troubles common with the Henderson engine.

Swen Swanson's single-place SS-3, shown here at it's birthplace, the University of South Dakota Engineering Shops. There are many similarities to the first Parrakeet model. The engine is a two cylinder Lawrance. UNIVERSITY OF SOUTH DAKOTA

Jack Rose and his original Parrakeet. Note the underslung wing. BARRY TAYLOR

Other than the engine problems, the plane flew well, so Rose searched for a larger and more reliable powerplant. He was fortunate to obtain one of the very early production Continental A-40's, and from the first flight he knew it was the motor he needed. He describes the performance:

"When Continental came out with their 37 HP model, I bought one of their originals, made a quick installation with practically no emphasis on a decent cowling job, and it flew like a tornado. I recall a group from Continental came to the airport by car during the winter when we had plenty of snow on the field; much too deep they thought to fly such a small ship. Well, it flew fine, and to put it mildly, seeing those white collar brass running around in the snow up to their knees like a bunch of crazy kids would encourage the most optimistic."

Accordingly, a second model was built, perhaps using some of the components of the first, powered by the same Continental A-40, probably an A-40-2. It contained a number of noticeable modifications. The lower wing was no longer underslung, the center section pylon was replaced with a conventional N-type, a pleasing engine cowling had been fashioned, a single strut replaced the flying wires, and the landing gear was lightened.

This plane, Rose Parrakeet model A-1, S/N 101, received registration number X13677 in the latter part of 1933. It flew exceptionally well, so with further encouragement from local interests, Rose decided to obtain an Approved Type Certificate, needed if he were to produce planes for commercial sale.

Bill Steiglitz, an engineering graduate of MIT, was hired to make the mechanical drawings and do stress analysis calculations. This project was supposed to take three months, but instead turned out to be a three year long series of frustrations for Jack Rose. He became more and more impatient; his relations with the local CAA and the design engineer were decidedly cool. He is quoted as saying *"I thought the engineer would never get the damned stress analysis done"*.

One example was the engine location. When originally flown, the plane was quite tail-heavy. Both Steiglitz and Rose tried to calculate the changes needed, but they could not agree. Finally they piled sheet lead on the top of the engine until the plane was in balance, then weighed the lead. A simple center of gravity calculation showed the engine should be moved 4 1/2 inches forward. Real barnyard engineering, but effective.

A second plane, S/N 102, was built in the latter part of 1934, receiving registration number X14842. But it was a year later, 8-28-35, that it was granted Group 2 Approval #2-514 and a "C" replaced the "X" on the registration. Finally, Jack Rose was in a position to actually produce planes for sale.

However, even though it was a fine little plane, one flaw in the design almost guaranteed it would not sell in any quantity. It was single place. You could not haul passengers in it, you couldn't give flying instruction in it, and you really could not go anywhere in it. Since you couldn't make money with it, it really became a rich man's toy, and not many could afford thousand dollar toys in the 1930's.

Side view of X-13677 JUPTNER

S/N 2, NC-14842 with the Scott engine installed BARRY TAYLOR

A modest advertising campaign was launched, (See full page ad following), and a few sales were made starting in 1936. This production prototype, S/N 102, was sold without an engine to use as a test bed for a new powerplant being developed by Leland L. Scott of Kansas City.

Scott had developed a two cylinder, two cycle air cooled aircraft engine, and hoped to sell it to the aircraft trade. It claimed 40 horsepower, had battery ignition, and drove a Hamilton metal propeller. He installed his engine on the Parrakeet, it looked much like an A-40 installation, and made a number of flights from his home base at Fairfax airport in Kansas City, Kansas.

Harold Maloy remembers seeing the plane flying over his high school in north Kansas City sometime in 1936 or 1937. *"It sounded just like a big model airplane motor, high RPM and lots of noise. Scott flew around our school several times, and of course I had to go out and look. The plane wasn't very high, and the motor didn't seem to be giving him much power. I think he only made one of his motors."*

This plane was sold to a Minnesota man in 1944, still retaining the Scott motor. Later, in 1949, it was again sold, less engine, to Howard Libersky of Mason City, Iowa. The airframe was strengthened, a Lycoming O-235 engine was installed, and the plane became a familiar sight on the national air show circuit.

For a time it seemed that the Rose enterprise would be successful. Two more planes, S/Ns 103 and 104, were sold in the summer of 1936 to west coast buyers, and Jack took the unusual step of delivering them himself, but not by flying. He made a simple trailer out of a Ford chassis, loaded on the dismantled planes, and pulled the rig to California behind his Model A Ford. Desperate selling measures such as these were needed during those tough depression times.

One thing the Parrakeet had going for it was its easy flying habits. This was demonstrated to Buck Hilbert one day when he was visiting the River road Airport near Old Orchard, Illinois sometime in the late 1930'S. He relates this experience:

"I was wandering around the field when I noticed a neat single place biplane sitting outside the hangar, and fussing with it was a tall, good-looking guy named Jack Rose. I struck up a conversation, asking all sorts of questions, and got some interesting answers. He asked me about my flying and I mentioned I had about eight of solo time in a Cub, Chief and Porterfield. Then I asked Mr.Rose how one learned to fly in a single place airplane. He looked at me and said something like 'You just do it'. 'How', I asked. 'Climb in', he said. And I did. Boy what a wonderful feeling. I was five foot six then and weighed about 127 pounds. The plane was just built for me!

"The next thing I knew, Mr. Rose was hollering, "Contact", and the A-40 was running. All the while I was taxiing out I was right on the edge of the seat and

Jack Rose wrote on the back of this photo: "Stopping to cool off on the way to California with 14843 and 44." UNDERWOOD

ROSE PARRAKEET MODEL A-1
$ 1,475.00 Flyaway Chicago

TO THE PRIVATE FLYER

An aeroplane which is not a power glider but a small ship, groundworthy as well as air worthy, with an operating economy within reach of the average man. Compare the size and note the saving in housing alone. The quality of the materials as well as the workmanship would normally put the selling price at $2000.00. We are offering this ship direct to the buyer at a price, with a saving of a dealer's commission and a minimum of profit to our selves, believing that the demand for a ship that is within reach of the average man to operate once he owns it will create a demand that will allow a production that will pay us a reasonable return for the time spent and the money invested.

GUARANTEE

The plane is Government Approved and bears an NC license, and we guarantee that any money deposited with us with order will be immediately refunded if the purchaser is a qualified pilot and is not fully satisfied after flying his plane.

REFERENCE

Lake Shore Trust & Savings Bank, Chicago, Ill.
Any Parrakeet Owner

SERVICE

We guarantee replacement parts at the lowest possible cost to owners.

PRICE

$1,475.00 Flyaway Chicago. Convenient terms to responsible people, $975.00 less Engine.
Any Federal or State tax is to be paid by purchaser.

SPECIFICATIONS

One place open Bi-plane, confirming to small plane design not a light plane.

Power, 37 H. P. Motor. Continental A-40-3.

Span 20', Length overall 16'-4", Height 5'-8".

Wing Area 116 sq. ft.

Wing loading 6.2 lbs. per sq. ft.

Power loading 19.6 lbs. per h. p.

Weight empty 470 lbs.

Useful load 258 lbs.

Gross weight 728 lbs.

Fuel capacity 10 gals.

Oil capacity 1 gal.

Maximum speed with 37 h. p. 100 M.P.H. Cruising Speed 85 M.P.H.

Service ceiling 12,000 feet.

Rate of climb 750 feet per minute. Cruising range 340 miles.

The ship takes off and lands in a space of 150 feet. It's landing speed is 35 M.P.H. in still air.

Fusilage is tubing and welded, fared with spruce, and treated with Berry Bros. Zinc Cromate Primer to prevent corrosion and rust. The Motor mount is detachable. The covering is of Grade A Flightex Fabric treated with dope and finished with Berry Bros. Pigment and is hand rubbed to a high luster. The motor coweling and cockpit cowels are of aluminum and finished with the same colors as the fusilage.

The wings are made up comprising of an extra select grade of spruce spars, inner braced with a system of tie rods, and a double system of compression members. The ribs are spruce and plywood glued and nailed and greatly exceeds the Bureaus requirements, and are guaranteed to hold a true wing section tho the fabric is tight. Tho their strength exceeds the requirement by margin of over 100% their weight is exceptionally low. The trailing edge of the wings are formed from aluminum in a way to insure a firm and true leading edge. The ailerons are formed from spruce and plywood and acuated by an internal torque tube. Horns do not appear on the surfaces. The wings and ailerons are covered with grade "A" Fabric and doped and finished the same as the fusilage. The cabanne is braced with two stream line tie rods as well as the tail group. These are the only wires appearing on the ship as bracing.

The tail group is steel tubing, welded and treated with Berry Bros. Zinc Chromate primer and fabric covered. The landing gear is of the split axel type, using Rose Model D wheels of approved design and Goodyear Tires.

Standard equipment includes Gardner Wood propeller, hand fire extinguisher, switch, windshield, streamline head rest, safety belt, propeller spinner, Arnes engine controls and engine tools.

The instruments include altimeter, tachometer, oil pressure gauge, oil temperature, as well as gasoline indicator and engine primer.

ROSE AIRPLANE CORPORATION

4327 N. HARLEM AVENUE - CHICAGO, ILLINOIS - PHONE AVENUE 2393

Full Page Parrakeet Ad in Aero Digest, 1937

ROSE Parrakeet Model A-1

Rose Aeroplane and Motor Company, Chicago, Illinois

• President and General Manager: J. W. Rose. Vice-president: Earl Howe. Sales Manager: Ray Applegate. Chief Engineer: William I. Stieglitz.

One-place open biplane. Continental A-40 engine, 37 horsepower. Span 20 feet. Length overall 17 feet 6 inches. Height overall 5 feet 8 inches. Wing area 116 square feet. Power loading 18.2 pounds per horsepower. Wing loading 5.8 pounds per square foot. Empty weight 410 pounds. Useful load 265 pounds. Gross weight 675 pounds. Fuel capacity 10 gallons. Oil capacity 1 gallon. Maximum speed 100 miles per hour. Cruising speed 85 miles per hour. Landing speed 30 miles per hour. Service ceiling 12,000 feet. Rate of climb 750 feet per minute. Cruising range 340 miles.

Fuselage: fabric covered and rubbed Berryloid high-luster finish; fairing is of spruce; welded steel tube framework, with a detachable engine mount. Wings: fabric covered; spars are solid spruce and are held by a double system of compression members and a single set of tie rods; ribs are of plywood and spruce and are glued and nailed; the trailing edge of the wing is of reinforced aluminum; ailerons are of spruce and plywood, glued and nailed and are actuated by a torque tube within the wing and are available only on the lower wing; interplane struts are of streamlined tubing; cabane bracing is by two streamline tie rods. Tail group: fabric covered; welded steel tubing; braced by single system of streamline tie rods. Split-axle type landing gear equipped with 16 × 4 Goodyear tires, and 16 × 4 aluminum disk wheels and cord shock absorbers.

Standard equipment includes Gardner wood propeller, hand fire extinguisher, switch, primer, windshield, streamline head rest, aluminum fuel tank engine cowl and propeller spinner, Arens engine controls, engine tools.

Instruments: altimeter, tachometer, oil pressure gauge, oil temperature gauge, fuel gauge.

Early Parrakeet Specs AERO DIGEST

just about as apprehensive as you can get. There wasn't much of a run-up. With single ignition, the gas was on so that was all there was to check. I lined up, opened the throttle and away we went.

"The little 40 horse flailed away, and after what seemed like a very short run, we were off the ground and climbing. I left the field and flew out west where I did a few practice turns and felt it out. As I remember, it stalled straight ahead and didn't give me any trouble. After a few more minutes I headed back to the field and entered a nice downwind pattern, did a 180 degree side approach and made a beautiful landing."

Mr. W. J. Viau bought S/N 106 on 8-23-36, and entered it in the Ruth Chatterton Air Derby. Unfortunately, the plane crashed on 8-30-36 one week later. But Mr. Viau must have been a determined man. He had the engine returned to the factory, and installed in another Parrakeet, S/N 105, which was delivered to him in 1937. After selling four planes in 1936, and one in 1937, production almost ceased. Continental, Lycoming, and Franklin were all getting ready to introduce their newer, higher horsepower 50 HP models; the days of the Continental A-40 were numbered.

To capture this market, Rose built S/N 107 in 1938, and used it as a test bed for the 50 HP Franklin 4AC-150. The certification tests were done in Kansas City; it was late in 1939 before all the reqirements of the CAA were met. This was the only Parrakeet designated as Model F-2. Evidently no sales resulted from this installation, so in 1940 the plane was refitted with a Continental A-40 and sold in the Chicago area.

These records would indicate that at least six A-40 powered Parrakeets were manufactured, but that S/N 105 and 106 used the same engine. In 1939 S/N 108 was built as a Model A-4-C to use a 65 HP Continental but certification was not received before World War II put an end to most civilian aircraft production.

Menasco 50 HP installed in a Parrakeet BARRY TAYLOR

Doug Rhinehart's restored Parrakeet, NC13676 at the Blakesburg Fly-In BARRY TAYLOR

After the war, Rose sold the manufacturing rights to a Foster Hanniford, who started advertising and assembling aircraft from leftover parts. Several were sold under the name of "Hanniford Bee", and later plans were offered to homebuilders. It is uncertain how many planes were built from these plans. By the early 1950's, all factory production had ceased.

The Author's first contact with a Rose Parrakeet came sometime in the early 1960's when Doug Rhinehart brought his restored NC13676 to the AAA Fly-In at Blakesburg, Iowa. He had made the long trip from Farmington, NM to show off the little plane and explain his plans for putting them back in production .

Doug had always admired the little Rose biplanes, since he saw one at Springfield Missouri back in the late 1930's. So when he found a derelict languishing in a Demming, New Mexico hangar sometime after WWII, he bought it and restored it to flying condition. It became an immediate hit on the Fly-In circuit, even winning Grand Champion Antique at the big Ottumwa Antique Airplane Association meet.

And once he started collecting Parrakeets, it was like eating peanuts, it was hard to stop. He ended up with four, and at that point, began to plan the production of new planes, this time with 90HP engines instead of the tiny A-40. He believed the ship had real potential as an aerobatic sportplane. In his search for help and information, he struck up a close friendship with Jack Rose, who by this time had retired. This eventually led to an agreement whereby Rhinehart would build five new Parrakeets; one without an engine would be Jack's license fee. Doug did complete four planes, but was unfortunately killed in a flying accident, so the project was discontinued, thus bringing to an end the Rose Parrakeet story.

Jack Rose and Doug Rhinehart in 1966 BARRY TAYLOR

Specifications

Rose Parakeet
ATC # 2-514, 8-28-35

ROSE AEROPLANE & MOTOR COMPANY

MODELS: Rose Parakeet A-1, 1 POLB

T.C. NUMBER: 2-514

Engine	Continental A-40, A-40-2, A-40-3 or A-40-4
Placard limits	For all operations 2575 rpm (40 hp)
Fuel	10 gallons (one tank in fuselage)
Cil	1 gallon
No. passengers	None
Baggage	None
Standard weight	728 lbs.
C.G. limits	(+0.5) and (+0.75)
Specification basis	Aero Bulletin 7A, Section 3
Serial numbers	101 and up eligible (See NOTE A)

QUIPMENT: (Datum is lower wing L.E.) (* Means net increase)

Class I:
- 101. 14x3 wheels (Rose model D) with 16x4 tires
- 102. 10 strands of 5/8-inch shock cord on shock strut
- 103. Tail skid
- 104. Propeller - wood (fixed pitch)

Class III:
- 301. 3-inch wheels (Goodyear 3TW) 3 lbs.*

NOTE A. Each aircraft certificated must satisfactorily pass:
- (a) An inspection for workmanship, materials and conformity before any covering, metal priming or final finish is applied. All woodwork may be varnished.
- (b) A final inspection of the completed aircraft.
- (c) Check of flight characteristics.

Chapter Nine

Taylor (Piper) J-2

A very early Taylor J-2; the N-number would indicate it was assigned in 1935. The prototype built in October of 1935 was NX15931, S/N 500. WHITTIER

In any history of the Continental A-40, the Taylor, later Piper, J-2 certainly deserves a place of honor, if for no other reason than it was the most prolific. More than 1200 J-2's were manufactured during the years 1936-38. But the J-2 has a sad legacy also. It was, at least indirectly, responsible for the contentious parting of Taylor and Piper.

The Taylor E-2 had served the Taylor-Piper organization well, but after over four years of production, changes were needed, changes in appearance and changes in performance. The E-2, for all its fine flying qualities, was SLOW! It would not cruise over 65 MPH, so any headwind slowed you down to a crawl. And it was beginning to look like an antique. The squared off tail and wings belonged to another era, as did the open cockpit.

Thus in the fall of 1935, Piper gave permission for a redesign, directing that the plane be modestly upgraded. But to keep costs down, no major structural modifications would be allowed. With only minor changes, it was hoped a new ATC certification, with the resultant CAA involvement and delays, could perhaps be avoided.

With these guidelines, Ray Carlson, assisted by Walter Jamouneau, came up with several cosmetic and aerodynamic modifications. First the turtledeck was raised to the level of the wing trailing edge, achieving a true cabin arrangement. Next, the wing tips, elevators, and rudder were given a pleasing rounded shape. Perhaps most important, the landing gear was given a wider tread, from 48" to 72", and moved 5" forward. This made the plane much easier to control in crosswinds, and helped to prevent noseovers.

Even though no basic structural changes had been made, the engineers at the CAA decided that a new stress analysis would need to be submitted. This caused weeks of delays, and acrimonious exchanges between Piper and Taylor, who was home ill most of the time, bedridden and recovering from a serious operation.

Much has been made of their differences regarding the redesign of the Cub, which resulted in this new model, the J-2. Actually, the rift had been growing for some time. As Taylor saw it, Piper had been usurping authority over the engineering department during his absence. Piper believed there was a lack of management supervision in that area. Their relationship deteriorated, finally assuming crisis proportions just as the J-2 Cub was being readied for production.

This controversy has led to some interesting legends concerning the J-2 which seem to have little basis in fact.

First, some have written that Taylor was not involved in the redesigning effort, knew nothing about it, and was furious when he saw the first test model. This could not be true, since Taylor's letters at that time refer to the slowness of the J-2's ATC process, not the redesign itself. To quote his letter of 11-29-35 to Bill Piper: *"As the Drafting Department of the Taylor Aircraft Company has been working about two months so far on the change-over on the Taylor "Cub", and have made such poor progress, I would advise that the stress analysis be sent to Professor Newell, of M. I. T. or other competent person to have the old analysis checked and have a new one made if necessary".* This shows that Taylor was at least somewhat familiar with the redesign work. (It also shows Taylor's contempt for engineers. He always called them "draftsmen", not wanting to dignify their position.)

The first J-2, S/N 500, was completed in October of 1935 and given identification number X15931. Living in a home next to the airport, it seemed certain that Taylor would have seen the plane numerous times.

A second myth: Taylor was supposed to have "fired" the entire engineering department when he saw the new plane, charging they had made drastic changes without his permission. This seems highly unlikely. The letter above indicates he was unhappy with the slowness of their ATC calculations, not with the design itself. In fact, Ray Carlson, who was head of engineering at that time, joined Taylor a month or so later in Butler, Pennsylvania, when Taylorcraft was started.

Third, as a reward for Jamouneau's taking over the engineering department at that time, Piper named the plane a J-2, using his "J" initial. The rational numbering system used by the Taylor organization at that time, refutes this, as is shown by the chart at the end of this chapter. The time-line doesn't fit either. The J-2 designation was being used as early as October.

This J-2, X16395, was used to certify the J-2 on floats. While the performance was marginal, it would get off the water with two people aboard.

WHITTIER

The Taylor plant at Bradford, just before the 1937 fire. Numerous sheds had been added to the original hangar. TAYLOR

As fall turned to winter in Bradford, the Piper Taylor relationship steadily deteriorated. By December, 1935 they were not speaking, instead sending letters back and forth, sometimes several in one day. Both did agree, however, that things could not go on as they were. Finally, on December 18, 1935, C. G. Taylor and Bill Piper signed a five page document ending their troubled, debt-ridden partnership. For $4750, to be paid out over three years, Piper became sole owner. As told in the next chapter, Taylor would move to Butler, PA and form his new Taylorcraft company.

Regardless of the controversy surrounding its introduction, the Taylor J-2 was a fine plane, much improved over the original E-2. It was faster than the E-2 and more appealing in its appearance. Due largely to the cleaned up turtledeck-cabin arrangement, which smoothed the airflow over the fuselage and tail, the cruise was increased at least 5 MPH. Not much, but it was certainly welcome to Cub pilots on cross-country flights. The rate of climb and gliding angle also improved. The new landing gear made the plane much more stable, and ground handling easier when windy. Wheels with roller bearings replaced the crude brass-bearings in the E-2 wheels, which wore out prematurely in regular use.

From the customer's standpoint, the greatest attraction was the sleek new look of the plane. It had a much more modern appearance. This NEW look was emphasized in factory literature, as shown on the page following. The word NEW is mentioned no less than ten times!

After the certification difficulties were resolved, sales turned surprisingly brisk for the cold winter season. By the middle of March, 1936, the order backlog had reached 70, so a night shift was started to increase production. Plans were put in motion to have Air Associates Inc. assemble Cubs in their Long Beach, CA plant By the end of the year production had reached the rate of 25 planes per week and 165 workers were employed. Unit sales for 1936 were 515, the highest number of commercial planes ever built in one year by a single company.

But disaster struck on the night of March 16, 1937. As a night shift workman drilled a hole in a firewall, a spark fell on some dope-soaked debris, which ignited instantly. A huge fire developed, fed by the plants wood and paint supplies. In less than an hour, almost the entire plant was destroyed. Just when poised for another year of increased sales, and profits, Piper was out of business

ELEVEN TO SEVENTY -- ALL FLY THE NEW TAYLOR CUB

You can own a new "CUB" as easily as you can own a car

THE NEW "CUB" is America's first big safe airplane at pleasure car price. It costs less to fly than it does to run a popular priced car. And it is easier for the average person to fly than it is to drive a car. The New "CUB" is the outcome of hundreds of thousands of successful flights and millions of miles of flying. No effort has been spared to make it the safest, soundest ship in America – at a price that all can afford.

It is economical to operate-using but three gallons of gas per hour and little or no oil. It has a cruising speed of 70 miles per hour yet it will land safely in a small area. It will take you anywhere safely and bring you back safely. The great American urge to go places and do things is exactly met by the NEW "CUB". It is in tune with the times in both appearance and performance. Still another safety feature is provided in the dependable NEW "CUB" Power Unit – a Continental Motor with all of the inherent smoothness, power and safety afforded by four cylinders. This means that if one cylinder fails in flight you lose only 25% of the power of the NEW "CUB" against 33–1/3 or 50% in three and two cylinder motors.

The youngest "CUB" Pilot on record is 11 years of age and the oldest is so sensitive to the age question that he won't tell. Whether you are eleven or seventy, man or woman, experienced pilot or just air minded, see and fly the NEW "CUB" America's SAFE Airplane.

This factory literature lauds the many changes made in the NEW Cub. Note they mention the safety of a 4-cylinder power unit vs. a two-cylinder. An obvious reference to the Aeronca C-3, one of their chief competitors. HILL

The Bradford Era newspaper reported this shocking story on the front page of its March 17, 1937 issue.

"The plant of the Taylor Aircraft Company was almost destroyed in less than two hours last night by mad, billowing flames that swept completely through the frame and metal sections of the factory with unbelievable speed...

"Firemen, called just before 10:45 P.M., were confronted with flames shooting high above the roofs of the place on their arrival. It appeared for some time that the metal building, the main sector of the plant, could be saved, but a strong and shifting wind aided the fire.

"Flames shot at least 50 feet in the air as one explosion followed another to various parts of the plant, which spread over considerable ground adjacent to the Harri Emery airport in West Branch. Employees said the blasts were from five-gallon cans of 'dope', liquid celluloid paint used to treat the fuselages in the finishing process. About 20 employees, working on the night shift when the fire started, wheeled 15 planes, most of them almost completed, onto the snow-covered landing field."

Piper had received the news while attending an aviation event in Los Angeles. He spent four and a half long days driving home, all the while pondering how his company could be saved. The plant carried almost no insurance, so the loss would be nearly complete. The crew at the factory had done what they could. Men were working on the planes saved from the fire, readying them for delivery. At least this would furnish some cash income.

The news of the fire was immediately noted by the officials of the Taylorcraft factory, now in Alliance, Ohio. On March 20, 1937 the Alliance Review newspaper reported huge expansion plans.

Need More Factory Space To Meet Rush Orders For Product

Taylorcraft Aviation Company Flooded With Demand Sales

"AIRPLANE OF DAY" SAYS CAPT. HAWKES

Taylor Cub Plant Disaster Leaves Taylorcraft Alone In Field

Flooded with orders for the monoplane that has proven a sensation in airplane manufacturing circles since its introduction a few months ago, the Taylorcraft Aviation Company plans immediate construction of an addition to the plant north of Alliance.

Fifty per cent more floor space must be provided to meet production demands, President C. G. Taylor said Saturday. A fireproof one-story addition is planned for the east side of the present building and a two-story addition in front, facing the North Benton Road.

The one-story addition would house the paint and dope shop. The two-story expansion would furnish more factory space on the first floor and office quarters upstairs. These additions would give [illegible] square feet more space than is now available and would enable the company to produce airplanes at the rate of at least [illegible] a week, Taylor said.

Expectations are that this expansion program will get under way about April 1. Details are being worked out, F. L. Wehrhan of New York City, a company director, having been in the city in that connection Friday. Before the work can be started, certain details, including purchase of the field from the Alliance Aircraft Corporation, must be handled. City Council cleared the way for the land purchase by cancelling the city's lease on the tract.

"Swamped" With Orders

The flow of orders into the Taylorcraft office has been little short of phenomenal, officials said. Sixty new orders arrived Friday, following destruction by fire of the Taylor Cub plant at Bradford, Pa. These orders, including [illegible] from a single firm in Iowa, boosted the total booked orders of the company to more than [illegible].

Following the Taylor Cub fire, the Taylorcraft office was flooded with queries from jobbers in scattered points of the United States seeking the right to represent the Alliance company in their territory. Telegrams and telephone calls arrived from such distant points as Louisiana and the Carolinas.

To Increase Force

Taylorcraft has about 60 employes on the payroll and will boost this number by at least a dozen Monday when that number of skilled Taylor Cub workmen will be given employment. These men have been engaged since the Bradford fire.

This March 20, 1937 Alliance Reveiw news article shows that the Taylor executives were eager to take advantage of the Piper misfortune. They were even pirating the work force.

Competitors believed Piper was completely out of business. However, the Cub managers were a resourceful group. Factory space was being sought throughout Bradford. Sales and engineering offices were set up in the airport administration building. An old abandoned iron foundry became the machine shop, and the large main airport hangar was used for final assembly. New blueprints were hastily obtained from the files at the U. S. Department of Commerce.

The Air Associates facilities on the west coast were even utilized. Wing parts made in Bradford were shipped west, while welded fuselages and other components were shipped east. Twenty two complete planes were made in Long Beach during this period, for which the CAA required a new ATC. Thus, in a rather unusual move, ATC #620 was obtained for these "Western Cubs".

But these were only stop-gap efforts at best. It was soon evident to Bill Piper that unless a suitable plant building could be found, his company could not survive. Actually, this search for a new site proved surprisingly easy. In 1937 the depression was still in full swing. Cities throughout the country would do almost anything to attract a plant with a payroll. As word of Piper's fire circulated, offers to help in relocation came in from across the nation. Many offered generous inducements such as buildings, bank credits and tax incentives.

Purchase of Land First Local Step; Building Deal On

Taylor Co. Heads Expected Here To morrow to Discuss Terms of Leasing Proposed Municipal Airport, to Cost $100,000 to $125,000 as WPA Project

EXPECT SHIFT OF ACTUAL OPERATIONS VERY SHORTLY

The Taylor Aircraft Co. of Bradford may be making airplanes in Lock Haven within the space of a few weeks.

This is the confident prediction of local civic and municipal leaders after a conference yesterday, at which a plan was evolved to provide a modern, municipal airport to be leased to the Taylor Company when they carry out their plan to move their manufacturing plant into the big, fireproof building of the Susquehanna Silk Mills, at the eastern edge of the city.

The conference, in which the City Council, the County Commissioners and the Board of Directors of the Chamber of Commerce took part, was followed by a telephone conversation last evening between E. F. Heffner, who has represented the city in the negotiations to bring the Taylor plant here, and Theodore V. Weld, president and sales manager of the firm, at Bradford.

This April 23, 1937, news article details plans for the Piper move to Lock Haven. LOCK HAVEN EXPRESS

In April, urged on by one of his distributors from that city, Piper chose Lock Haven, PA, which had offered him a huge abandoned silk mill at a bargain price. They also obtained a WPA grant of $125,000 to build a new airport adjacent to the factory. The move was made in June, and Piper was back in business. Most of the 165 employees came also.

This location offered a number of advantages to Piper. Most important was the huge two-story manufacturing plant, bought for about 10 cents on the dollar. The building, with unencumbered floor space, would allow a true assembly line to be installed. There was a rail siding adjacent, making shipping and receiving convenient and inexpensive. The new landing field would be larger and better suited for aircraft testing and delivery. An infusion of $125,000 federal WPA dollars would benefit the entire community.

Although not mentioned in news articles, the labor situation would favor Taylor also. There had been grumblings in Bradford, a high-wage oil town, about Piper's 25 cents per hour pay rate. But the unemployed factory workers in Lock Haven would not be as likely to complain.

With the new facilities, the production capacity was greatly increased; even after the delays caused by the fire and moving, 515 planes were made in 1937. The spacious new plant allowed for greatly increased production, and the economic conditions in the country had begun, ever so slowly, to improve.

A new name came with the move. Not wanting to help advertise the Taylor name any longer, especially with the advent of the Taylorcraft, the Lock Haven company became the Piper Aircraft Corporation. The last 61 J-2's would carry "Piper Cub" nameplates.

With the introduction of the A-40-5 twin ignition engine, Cubs and other American lightplanes became eligible for export to several European countries. One such export was S/N 1166 which , along with three other Cubs, went to England in 1938. Equipped with wheels, skis

G-AFFH in 1938 HONEYCHURCH

Here's a Plane that *You* Can Own!

NOW, at last, it is easy to afford the pleasure of flying one's own plane! For the New Silver Cub places what used to be the sport of millionaires within the reach of everyone. At its remarkably low price of $1270, F.A.F. factory—only $425 down and the balance in easy monthly instalments—the New Silver Cub costs no more than a medium priced car!

Think of the enjoyment you can have with this safe, performing airplane! For week-end trips to distant cities—for vacation jaunts cross-country—or for cool flights on hot summer days—the New Silver Cub is ideal!

Its smooth-running, 4-cylinder motor gives the Cub utmost safety, dependability and economy. Owners everywhere report gas consumption of less than 3 gallons per hour and almost no oil! Easy to buy and easy to fly, the New Silver Cub is the world's fastest selling airplane. Order yours now at the amazingly low price of $1270. (Prices are subject to change without notice.)

FREE! Send for our handsome, new folder illustrating and describing the New Silver Cub and for the name of your nearest Cub dealer today. Paste coupon at the right on penny post card, if you wish.

TAYLOR AIRCRAFT COMPANY
77 P Street, Lock Haven, Pa., U. S. A.
Cable Address: CUB

TAYLOR AIRCRAFT COMPANY
77 P Street, Lock Haven, Pa., U. S. A.

Please send me your new, illustrated, free folder and name of my nearest Cub dealer. No obligation.

NAME ____________

ADDRESS ____________

CITY ____________ STATE ____________

The New Silver CUB

THE WORLD'S FASTEST SELLING AIRPLANE

A 1937 ad from Lock Haven

AERO DIGEST

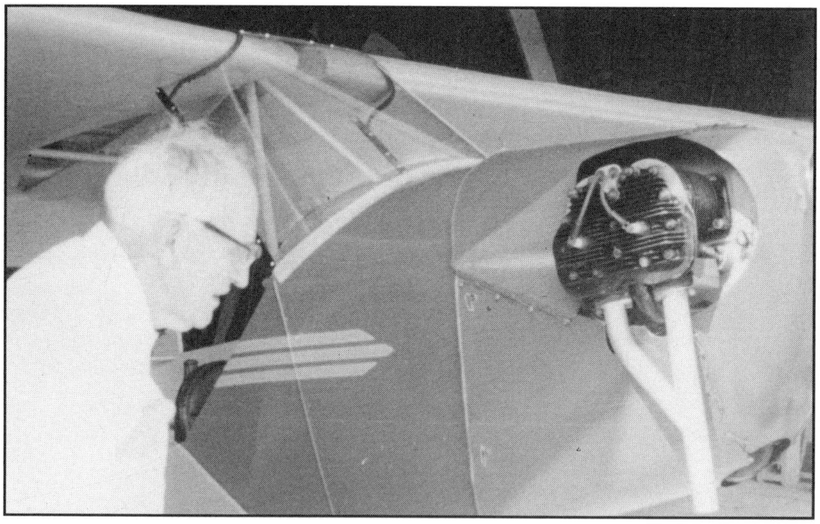

C. G. Taylor inspects a restored J-2 HILL

and Edo floats, it was given the British registration mark G-AFFH. Charles Gardiner, a famous British racing pilot, flew the little plane on floats around the Cornish coast that summer. In his journal he describes it as being grossly underpowered, barely getting off the water with two people, but still a joy to fly.

On a flight to Torquay he comments: *"At Torquay we crossed the point, which was the cause of a few nasty seconds on the leeward side, where there were quite considerable down-droughts. The Cub's 40 HP cannot cope with a very high rate of ft. per minute draught downwards, so we rapidly found ourselves getting unpleasantly close to the water. This is no cause for worry, however, since air cannot go down into water, and the downward gust must flatten out sometime."*

During WWII the plane, like most others, was put into storage, then after travels around England and Spain for 30 years, ended up in the capable hands of restorer Martin Honeychurch in Wiltshire. As shown in the photo on this page, Martin is well along in the rebuilding process.

As might be expected, the little J-2 has become a favorite among antique aircraft restorers. Dick and Jeannie Hill's pristine NC-17854 has been a familiar sight at the Blakesburg and Oshkosh fly-ins for many years. In 1980 the Hills hosted a mini fly-in at their Harvard, IL airstrip, attended by several J-2's, an E-2, a Taylorcraft Model A, and C. G. Taylor himself! Dick recounts the event:

"That evening, Mr. Taylor was sitting on the ramp among six of his planes in front of him, watching the guys give rides. He said, "There was never this much activity at Bradford!"

With years of experience flying A-40 engines, Dick Hill has had remarkably little trouble. He reports:

"After flying the E-2 and J-2 for more than thirty years we have had three forced landings and two of them were my fault. The only time when I was really let down was when a coil failed. I had taken off, made a turn back over the field and was passing along side the runway about 200 feet when it quit, went from 2400 to a dead stop. I made a 180 and landed long. With a tailskid and no brakes I slid through the end of the strip and about 100 feet into a field of recently husked corn. Luckily, no damage".

In 1938, the first of several 50 HP engines were appearing on the scene, led by the Continental 50 installed in the Luscombe Model 50. This meant a redesign of the Piper J-2 would be mandatory. The new model would become the famous Piper J-3. Its story will be told in Chapter 14.

In all, 1207 J-2 Cubs were built during the period beginning in March of 1936 until October of 1938; NC20175, S/N 1975, was the last.. More J-2 Cubs were produced than any other A-40 powered plane.

G-AFFH under restoration in 1999 HONEYCHURCH

TAYLOR Cub Model J2 2-place **ATC 595**

Specifications. Span 35′ 2.5″ (10.74 m.). Length 22′ 6″ (6.87 m.) Height 6′ 8″ (2.03 m.). Wing area 178.5 ft² (16.6 m.²). Power loading 24.2 lbs./h.p. (10.95 kg./h.p.). Wing loading 5.42 lbs./ft.² (26.6 kg./m.²). Weight empty 563 lbs. (255 kg.). Useful load 407 lbs. (184 kg.). Payload 175 lbs. (79.4 kg.). Gross weight 970 lbs. (440 kg.). Baggage 5 lbs. (2.3 kg.) Fuel 9 gals. (34.1 lit.). Oil 1 gal. (3.8 lit.).
Engine. Continental A-40-4, 40 h.p. at 2575 r.p.m.
Performance. Maximum speed 85 m.p.h. (137 k.p.h.). Cruising speed 72 m.p.h. (116 k.p.h.). Landing speed 30 m.p.h. (48.3 k.p.h.). Service ceiling 12,000 ft. (3657.6 m.). Climb per min. 450 ft. (137.2 m.). Cruising range 210 mi. (338 km.).
Construction. Fuselage: Warren truss frame of chrome molybdenum and carbon steel tubing; detachable engine mount; steel fairing strips; tandem seating arrangement; dual controls; quick opening section in fuselage to right of seats to permit easy entrance and exit; fabric covered. Wing: U.S.A. 35B airfoil section; solid spruce spars; steel tube compression struts; steel tie rods; riveted aluminum alloy ribs; Vee type external bracing consisting of two streamline tubes meeting at lower longerons; fabric covered. Tail group: welded steel tube and channel sections; wire bracing of chrome vanadium tie rods between rear spars of surfaces and lower longeron; stabilizer adjustable in flight; fabric covered. Fixed split type landing gear with shock cord rings. Leaf spring tail skid. Licensed by Dept. of Commerce as a seaplane on Edo model D-1070 floats.
Standard equipment. Sensenich type 69-D wood propeller, Goodrich 8.00 × 4 tires, Hayes wheels, shock cord ring absorbers, Grinnel fire extinguisher, Johnson & Johnson first aid equipment, fabrikoid upholstery, Pyralin windshield and cabin enclosure.
Instruments. Aeromarine altimeter; U.S. Gauge oil pressure gauge and oil temperature gauge; Waltham tachometer; Taylor fuel gauge.

Published specifications AERO DIGEST

C. G. Taylor's model designation and serial number system

C.G. and Gordon Taylor started their serial (construction) numbering system with their first plane, the modified Curtiss Jenny, which received S/N 1, but no model number. The Loomis plane followed, then the Chummy which received the Model designation A-2. The "A" indicated the first model, the "2", seating capacity. This system was followed throughout the Cub production. When Taylor left Piper and started Taylorcraft, he used the same model system, first the Model A, then B etc. The chart below shows the progression.

Date	1st serial number	Model designation	Description
1926	1	None	Curtiss Jenny
4-27	2	None	Loomis Monoplane
2-28	3	A-2	First Chummy
6-28	6	B-2	Kinner Chummy
9-29	7	C-2	Giggenheim Plane
5-30	None	D-1	Glider
10-30	11	E-2	First Cub
2-34	66	F-2	Aeromarine Engine
10-34	149	G-2	Taylor Engine
5-35	?	H-2	Szekley Engine
	"I" designation not used, easily confused with "1"		
10-35	500	J-2	First J-2

Specifications
Taylor, (Piper), J-2 Cub
ATC #595, 2-14-36

PIPER J-2, 2 PO-CL-SM, ATC 595

Engine	Continental A-40, A-40-2, A-40-3, A-40-4 or A-40-5 (Dual ignition), (See Item 309)
Placard limits	(A-40) (A-40-2) (A-40-3) Maximum, except take-off — in. Hg., 2550 rpm (37 hp) Take-off (one minute) — in.Hg., 2550 rpm (37 hp) (A-40-4) Maximum, except take-off — in. Hg., 2575 rpm (40 hp) Take-off (one minute) — in.Hg., 2575 rpm (40 hp) (A-40-5) Maximum, except take-off — in. Hg., 2575 rpm (40 hp) Take-off (one minute) -- in.Hg., 2575 rpm (40 hp)
Propeller	Maximum permissible diameter 81 in.
Placard speeds	Not required
Fuel capacity	9 gals. (one in fuselage)
Oil capacity	1 gal. (in engine crankcase)
No. passengers	1 (+9) or (+36)
Baggage	(1 compt. aft rear seat) Landplane 20 lbs., Seaplane 12 lbs. (Placard compt.: "Includes anchor and rope 12 lbs. when carried.")
Weights	Empty Use actual (Seaplane approximately 88 lbs. net increase over landplane) Standard - 970 lbs. (Serials 500 thru 599) Landplane - 1000 lbs. (Serials 600 and up) Seaplane - 1068 lbs.
C.G. limits	Landplane (+11.8) and (+20.4) Seaplane (+10.2) and (+20.8)
Spec. basis	Approved Type Certificate No. 595
Serial numbers	Landplane - 500 and up mfrd. prior to 9-30-39 eligible. Seaplane - 658 and up mfrd. prior to 9-30-39 eligible. Approval expired as of that date.

EQUIPMENT: (Datum is wing leading edge)

Class I.

(a) Landplane

- 101. 7.00-4 wheels (Shinn) and tires
- 102. Tail skid — 3 lbs.
- 103. Wood propeller — 12 lbs.

(b) Seaplane: Item 103 PLUS

- 151. Edo D-1070 floats — 145 lbs.
- 152. Aux. fins (on left and right stabilizer)

Class II.

(a) Landplane

- 200. Miscellaneous items as noted in approved weight and balance report.
- 201. Everel propeller I-38-A or -B (single blade) — 12 lbs.*
- 202. Wheel streamlines — 6 lbs.*

(b) Seaplane - Items 200 and 201

Class III.

(a) Landplane

- 301. 18x8-3 wheels (Goodyear 3LNBM, ATC 26) — no change in weights
- 302. Position lights — 3 lbs.
- 303. Cockpit enclosure — 6 lbs.
- 304. Steerable tail wheel with 6.00-2 tire
  - (a) Aircraft Associates (Dwg. E-4) — 3 lbs.*
  - (b) Aero Activities (Dwg. 1-C) — 3 lbs.*
- 305. 8.00-4 wheels (Hayes model 840)
- 306. Battery (3-BRL-6) — 8 lbs.
- 307. Carburetor heater — 2 lbs.
  - (a) Model J2-A95
  - (b) Model J2-A902
  - (c) Poorman
- 308. Combination carburetor and cabin heater — 3 lbs.
  - (a) Model J2-L90
  - (b) Grieder
  - (c) Air Transport and Equipment (Model II)
- 309. A-40-5 engine with dual ignition — 13 lbs.*
- 310. Parachute - 1 or 2 at 20 lbs. ea.
- 311. Miscellaneous extra instruments
- 312. Grieder pressure cowling (Model A) — 2 lbs.
- 313. Skis
  - (a) Piper S-1000
  - (b) Air Transport A or 1220-480
  - (c) Federal SA-1
  - (d) Marston MFS-1200

(b) Seaplane - Items 302, 303, 307, 308, 309, 310, 311 and 312 PLUS

- 351. Edo 54-1140 float installation — No weight change (Item 152 unnecessary with this installation)

NOTE 1. Eligible for export as follows subject to inspection for equipment specified in Chapter XII of Inspection Handbook: (May 25, 1938)

(a) Canada:
1. Landplane
2. Seaplane provided modified front lift struts in accordance with Taylor Dwg. J2-151 are installed.
3. Skiplane - not eligible. However, structure complies with Canadian requirements for ski gear when item 313 (a) or equivalent is installed.

(b) Great Britain provided wing rear spars are reinforced in accordance with Taylor Dwg. J2-L15

(c) Australia as landplane only provided that:
1. Wing spars are reinforced in accordance with Taylor Dwg. J2-L15.
2. Modified front lift struts in accordance with Taylor Dwg. J2-151 are installed.
3. The following placard speeds are displayed:
   - Level flight or climb — 80 mph
   - Glide or dive — 108 mph

(d) All other countries.

Chapter Ten

Taylorcraft "A"

Test flight of the first Taylorcraft at Butler, PA TAYLOR

The Taylorcraft Model A was arguably the most beautiful and best flying airplane ever to use the Continental A-40 engine. And it was a product of that same genius in design, C. G. Taylor, the man who also built the famous "Cub".

When Taylor was forced to sell his interests and leave the Taylor Aircraft Co. in January of 1936, his future in the aircraft industry looked bleak. His only financial resource was the $200 a month payment he was receiving for his share of the Taylor Aircraft Company. But he did have friends in the aviation industry, and it would be one of these friends, Byron Shinn, who would help Taylor begin building planes again.

Shinn, an engineer, had started his Shinn Devices Company in the late 1920's, based in Butler, Pennsylvania. He specialized in making wheels and other accessories for the aircraft industry. Many of the lightplanes of the day such as the Cub, Aeronca, and American Eaglet, used Shinn wheels. They were marvels of simplicity, consisting of cast aluminum halves which bolted together to mount the commonly used 7:00X4 and 8:00X4 low pressure tires.

The Taylor company was one of Shinn's best customers, and over the years he and C. G. Taylor had become friends. So when he learned that Taylor had been forced out of the Cub organization, he urged him to stay in the aviation field, even offering to assist in starting a new venture.. He was able, through a Pittsburgh financier David Barr Peat, to raise sufficient funds for a new manufacturing company. It would manufacture a new plane designed by Taylor. In a 1992 interview Shinn remembered:

"We moved Taylor down from Bradford, rented him a house, and set up a small manufacturing facility in a vacant hangar. Shortly thereafter, several of Taylor's former employees from the Piper organization joined him".

The airport at Butler was also a product of David Barr Peat's promotional efforts on behalf of Pittsburgh area aviation. In addition to several gravel runways, it had two large brick and steel hangars with full opening doors at both ends. The Shinn Devices factory was located in a small building attached to one of the hangars.

TAYLORCRAFT

BY C. G. TAYLOR PITTSBURGH-BUTLER AIRPORT, BUTLER, PA.

STABLE **FAST**

ECONOMICAL A NEW LIGHT PLANE **LOW-PRICED**

First Taylorcraft ad, in Aero Digest, February, 1936.

According to Byron Shinn, the Taylors set up a drafting room in his small offices so that the design could be formalized and the stress calculations needed for the CAA could begin. It seems almost certain that Taylor had the design clearly in mind before he even left Bradford. He and Ray Carlson had been working evenings in Taylor's schoolhouse-home basement for some months. At one point, in the Fall of 1935, Piper had asked Ken Tibbits, the shop foreman, what Carlson and Taylor were doing. He had replied, *"They are designing a side-by-side plane."*

The first Taylorcraft advertisement appeared in the February, 1936, Aero Digest, a bit over a month after the Bradford "Split". This would indicate that Taylor had a new plane in mind before he left. He did not, however, have an airplane to show prospects or investors. Actually, even though the ads published in March and April purportedly showed the new Taylorcraft plane, it was really a model, carved out on his kitchen table by C. G. Taylor himself.

It is a tribute to C.G. Taylor's character and reputation that a number of the key employees at Bradford followed him to Butler, Pennsylvania. Ken Tibbits, the original "first employee" and shop foreman, came to help build the prototype and do the initial test flying. Ray Carlson, who had been Chief Engineer at Bradford until outranked by Jamouneau, came to work on the drawings and stress analysis. Harold White, another engineer, moved down to assist Carlson. Les Salchow, a Parks graduate and friend of Ken Tibbits, left his job building jigs for the J-2 Cub and began similar work for the new Taylorcraft.

These men must have felt an intense loyalty to Taylor. They left secure, responsible positions in the Piper organization for the obviously risky prospects of Taylor's new venture. Fortunately, all would be rewarded in time, advancing to high positions with Taylorcraft.

Arthur Taylor, C. G.'s father, moved down with the family and set up his machines in the borrowed hangar. While Arthur was a superb machinist, he had a gruff manner and was thought by some to be set in his ways, especially regarding shop practice. There was some friction with the Shinn personnel; the fact that they were "borrowing" space and tools from Shinn certainly contributed to a certain amount of discord. However, most agreed that without Arthur's machinery and skills the new Taylorcraft could never have been built.

Through the early months of 1936, construction went forward on the prototype but not as fast as Taylor had hoped. Only a small amount of money had actually been received from the investors, barely enough to buy an engine, materials for one plane, and living expenses for the factory crew. In 1983, Taylor told the Author, *"I built the first Cub in three weeks, but it took three months to build the first Taylorcraft.*

Shinn factory at Butler, Pennsylvania. SHINN

Taylorcraft model used for advertising TAYLOR

Mr. and Mr. Ken Scholter, 1991 AUTHOR

One reason was that I was continually travelling, trying to obtain additional financing. One day when I returned I found the crew had spent three days trying to design the engine mount. I hung the engine in place with a hoist, measured and cut the tubing, and in three hours we had the mount done."

The advertising campaign, which had begun almost as soon as the first financing was obtained, created an immediate demand for the new plane. Probably beyond their expectations, certainly beyond their ability to deliver. As soon as the price was announced in April, they began to take down payments for future deliveries. Robert Taylor, C. G.'s son, remembered one incident. *"The first Taylorcrafts were sold from a picture of a twelve inch wooden model my father made. One night, just after I'd gone to bed, Dad came up the stairs with ten $1,000 bills in his hand. He said they were deposits on the new Taylorcraft. To a young boy, this was very impressive."* In a June 1936 letter to the CAA, Taylor mentions having orders for 78 planes.

Ken Scholter, a local mechanic, was hired to do welding on the new plane's fuselage frame. Hiring Scholter had another fortunate consequence. He was an avid photographer, and the photos shown on these pages are the only record we have of the prototype Taylorcraft construction. Later, Mrs. Scholter was hired as secretary in the new firm's office.

Ken remembered cutting and fitting material until it "looked right", then having the draftsmen come out in the shop and take the measurements. In 1991, he and his wife still lived in their home, a half-mile from the hangar where the first Taylorcraft was built. Framed on the wall of his den is a letter of recommendation from Taylor dated July 9, 1936, the same day the firm moved to Alliance. The letter was quite congratulatory, but Scholter had this comment: *"They left without being able to pay our final wages, so the least they could do was give us a good letter!"*

As work on the first plane proceeded, publicity was needed to attract dealers and perhaps potential investors. The following press release went out to the aviation trade in April of 1936:

Butler, Pennsylvania

AMERICA'S NEWEST LIGHTPLANE

"Butler, Pennsylvania, April 8. With the industry demanding a truly modern lightplane with greater safety, economy, comfort and speed, to sell for less than $1500, Mr. C. G. Taylor, well known in aviation circles as the designer of the famous Taylor Cub, is presenting his latest achievement in the light plane field known as the Taylorcraft.

"Mr. Taylor has resigned his position as President and General Manager of the Taylor Aircraft Corporation, and has organized a new company which is located in the modern hangars of the Pittsburgh-Butler airport, where proper factory and field facilities are available and adequate capital provides an ideal setup."

The prototype Taylorcraft fuselage SCHOLTER

In the flowery language, (and long sentences), common in the press of that day, the release went on to describe the various mechanical features of the plane. It closed with this optimistic statement:

"Extensive equipment is being installed for the line production of the new ship, which Mr. Taylor personally designed and built. Dealer interest already indicates that Mr. Taylor will greatly exceed his last year's production, which represented over fifty per cent of the lightplanes in the United States."

As the first Taylorcraft took shape in the Butler hangar, it was evident that a new concept in lightplane design was being introduced. True, the basic dimensions and configuration of the new monoplane were similar to those of the Cub and Aeronca. But the comparison stopped there. Taylor had achieved significant improvements in utility, performance and appearance with his new design. With the same 40HP motor as the Cub, it was almost 20 MPH faster and certainly more comfortable.

The specifications and performance data were released to the press in April, before the first prototype had even flown. However, they would prove to be remarkably accurate. No significant changes were made to the data shown when ATC #643 was issued a year later on 6-11-37.

Not surprisingly, Taylor had reverted to the side-by-side seating arrangement of the Chummy. Taylor believed this facilitated communication between the pilot and passenger or student. It was also a clear imitation of some of the higher priced luxury cabin planes of that era.

Prototype nose cowl. Hand forming shows. Taylor

It had a pleasing, streamlined, almost racy appearance. Starting at the nose of the plane, the cowling was faired cleanly into the fuselage firewall and secured by clever twist fasteners instead of the finger-biting wire lacing of the Cub. Best of all, it was fabricated in two pieces, split horizontally so it could be removed without pulling the propeller.

First run-up of an A-40 in a Taylorcraft Scholter

It was not surprising that the Continental A-40 engine was chosen to power the new plane. Actually, there wasn't much choice. The A-40 was the only really reliable engine then made, and Taylor had experience with it in the Taylor Cub. Most of the A-40 production had gone into Cubs during 1931-35, so it is likely that Continental made some effort to see that Taylor's new plane used their motor. The engine was supported on a lightweight steel tube removable mount only an inch away from the firewall. Simple exhaust stacks extended down and out the bottom of the cowling through oblong openings. Thus the exhaust gasses were directed into the bottom slipstream, away from the cabin. This was a welcome change from the short stacks on the Cub, which let exhaust fumes and sparks fly directly into the pilots face.

The gas line ran directly from the ten-gallon cylindrical tank to a gascolator-strainer mounted on the Stromberg NAS-2 Carburetor. A built-in gas shut-off valve with a control knob on the dash, required two turns to open or close. The oil pressure line and the oil temperature capillary tube ran from their engine crankcase fittings through the firewall to their respective instruments. The only electrical wiring on the plane connected the push-pull switch on the dash to the magneto for grounding.

The landing gear was of tripod design, called the "Pursuit Design" in press releases, with the shock absorbing rings inside the fuselage for streamlining. This gave the plane a racy higher and wider stance. The wider stance made it more stable than the Cub, but visibility over the nose in a three-point attitude was minimal.

The wheels were a major innovation in the light plane industry. Taylor had convinced Goodrich to design a 6:00x6 tire so as to reduce the weight and help in streamlining. The tire company was interested in doing this because the 8:00x4 casing commonly in use was very hard to produce, given the technology of the times. The inner moulds for this tire could not be removed through the small center opening, resulting in the use of disposable Plaster of Paris forms. This was a time-consuming process that caused slower production and higher labor costs. The 6:00x6 tire eliminated this problem since it could be produced in conventional moulds. The size is still a standard for light aircraft today.

C. G. Taylor in 1936 TAYLOR

The Shinn Devices company, aided in this innovation by developing a new wheel that was a marvel of simplicity, light weight and low cost. The wheels were made of a cast inner half held to a pressed aluminum outer half by long through bolts, now using roller bearings. To keep the weight down, Taylor worked with Byron Shinn in developing a tapered axle arrangement, allowing a smaller outside bearing. It should be noted that Taylor was sharing space in the Shinn Factory at this time.

The tail skid arrangement of the prototype was less successful. It was mounted on a hinged steel frame, about thirty inches ahead of the tail post, much like the early Wacos. Rubber washers in compression acted as shock absorbers. This location worked fairly well on grass runways, but when a tailwheel was added for use on hard surfaces, the handling of the plane, especially when landing, became somewhat "sqirrelly". The earliest models had no brakes to help with this steering problem. Consequently, most of the earlier Model A Taylorcrafts were converted to the later design; a tail spring mounted on a bracket under the tail post, much like the Cub system.

The new plane's fuselage could be described as "fish shaped" with clean looking reverse curves from the back of the cabin to the tail post. It was fabricated of light seamless 1020 steel tubing, welded into a Warren truss design. An "X" of tension rods braced the frame above the passenger compartment. Thin strips of wood fairing gave a slightly rounded shape on all four sides.

A wide steel firewall separated the front of the cabin from the engine compartment. A cylindrical gasoline tank, holding ten U. S. gallons, was hung above the pilot's feet behind the boot cowl. This gave the plane at least a three hour endurance, adequate for its intended use. A wire and cork gas gauge indicated the fuel quantity. The windshield and side windows were of 0.050 DuPont Plasticine, fastened at the sides, top and bottom with sheet metal screws and rubber extrusions.

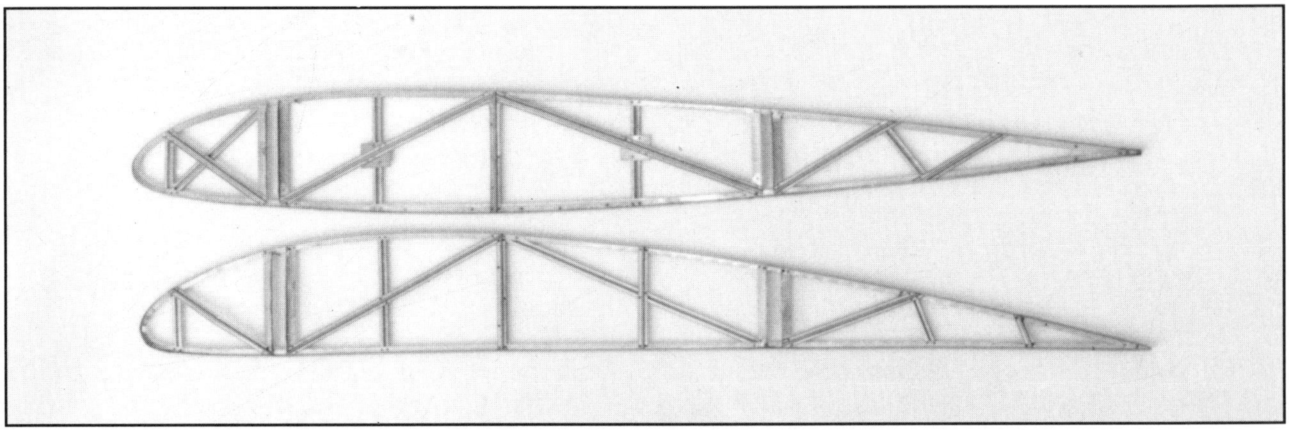

A comparison of a Taylorcraft wing rib, (top), with that of a Cub AUTHOR

The cabin, or "Cockpit" as it was still often called, resembled the appointments of the more expensive cabin planes of the day. The prototype used standard aircraft instruments, but production models sported a Stewart-Warner instrument cluster in the center of the panel housing a large six-inch tachometer as well as the oil pressure and temperature gauges. (This was the same instrument used on Dodge cars of that year.) The instrument might have saved the company money as well. A 1945 price list shows the entire instrument to cost $15.00 retail. On the pilot's side was an airspeed, 125 MPH maximum, and on the opposite side a non-sensitive altimeter. This completed the standard instrument layout, but an optional magnetic compass could be mounted on top of the dash.

A wheel control system, normally found on larger, more expensive planes, replaced the usual control sticks. This was a much more comfortable arrangement, especially if the passenger was a woman wearing a skirt. The round control wheels operated the ailerons and elevators, through a system of bicycle chains, pulleys and cables. Conventional rudder pedals were installed on the floor.

The wide seat, (39 inches), was a simple canvas sling, supported by three fuselage tubes. The center tube was moveable, so that by adjusting a lacing arrangement, the seat could be raised or lowered. Upholstered seat and back cushions rested on this sling, but could be removed when parachutes were being worn by the occupants. A small canvas baggage compartment, 6" x 9" x 39", was located behind the seat. It was placarded for eleven pounds maximum. The entire cabin was covered on the inside with the same doped fabric used for the rest of the plane.

A door was installed on the right hand side, with provisions for an optional left hand door if desired, extra cost, of course. This entry door caused one constant criticism of Taylorcraft airplanes. The door was quite small and required some near acrobatic movements to enter the cabin. The same sized door has persisted throughout the life of Taylorcraft, although later models sought to disguise the problem, making the door larger, but not the opening. On both sides, a large swing-out window allowed for ventilation.

The tail section was generous in size, and conventional in construction. Rounded curves were used on all panels, in contrast to the E-2 Cub's squared-off appearance. Round steel brace wires supported the panels. No trim mechanism was installed. Taylor believed that by locating the disposable load of fuel and passenger at the center of gravity, no trimming would be needed. This proved to be correct for the 40HP models, but when the higher horsepower engines were used in later designs, trim systems were installed.

At first glance, the Taylorcraft wing appears to be a carbon copy of the E-2 Cub. It has a 36 foot span, a 63 inch chord, with spar and rib locations nearly the same. The aerodynamic characteristics, however, were quite different. This was largely due to Taylor's selection of the relatively new NACA 23012 airfoil for his new ship. It provided substantially improved performance. A review of the airfoil profiles used in other 1936 lightplanes shows:

Taylor Cub	USA-35-B
Aeronca	Clark Y
Luscombe	NACA 4412
Waco	Clark Y
Stinson	Clark Y
Beech Staggerwing	NACA 23012

TAYLORCRAFT

By C. G. Taylor, Pittsburgh-Butler Airport, Butler, Pa.

First Taylorcraft Company Letterhead.

The 23012 airfoil offered good lift without the substantial drag associated with the USA-35-B and the Clark Y. It does have a discontinuity in the lift curve at the point of stall, but this was found to be less severe when the wing loadings were low. Rectangular wings resist tip stalls, and Taylor further insured a gentle and controllable stall situation by building a 3/4 degree washout twist into the wing structure.

In the photograph shown earlier, a comparison of the actual Taylorcraft and Cub wing ribs shows that while the Cub rib had an almost flat bottom, the Taylorcraft rib approaches a symmetrical shape with a pronounced convex bottom curve.

The wing spars were of clear spruce, slightly smaller in dimensions than those on the Cub. The rest of the wing hardware, compression members, drag and anti-drag wires, and attachment fittings, were conventional. The ailerons were of the improved Frieze design, which, while more difficult to manufacture, reduced the "aileron yaw" associated with the top hinged ailerons found on most planes built in that era, including the Cub. The wings were braced with streamline steel tube struts.

Throughout the Spring of 1936, Taylor and Carlson worked on the design of the new plane, then took the plans out to the shop for fabrication. Several of the Bradford people had come down to join the new company and work on the prototype. Arthur Taylor was busy building the needed jigs and fixtures, duplicating much of the machinery that had been left with the Cub organization.

The choice of engine was mainly dictated by availability. Taylor had worked through the A-40 troubles with the Continental engineers, it was now reliable, and relatively cheap. It had just enough horsepower to handle this new design, a two place plane weighing under a thousand pounds loaded. Adding to that, Taylor may have been able to have an engine loaned to him for this prototype. In early May, Continental A-40-2, S/N 635, was received and installed.

The initial funds raised by David Barr Peat were fast running out, so C. G. Taylor was continually travelling, looking for additional financiers. In later years he complained that progress slowed considerably when he was gone from the shop. But by the end of April, the structure was complete, ready to be covered with Grade A cotton fabric. Mrs. Taylor sewed the covers on her home sewing machine, then it was installed on the framework using clear nitrate dope. The plane was finished in a shiny "Gunmetal Grey", according to several sources.

Finally, the sleek little plane was finished and flight testing could begin. Ken Tibbits had the responsibility for making these flights, but he may not have been the first person to leave the ground in a Taylorcraft. Bob Taylor remembers riding with his father during some fast taxi tests, when either by accident or design, the plane became airborne for a few yards.

First Taylorcraft wing SHOLTER

The first Taylorcraft arrives at Alliance, Ohio ALLIANCE LIBRARY

At any rate, Tibbits was the person who did the bulk of the test flying, starting in mid-May. Byron Schinn remembers how he went about preparing for the first flight:

"He was very careful and cautious. First, he taxied up and down the grass runway, a little faster each time, testing the response of the controls. He then lifted the plane about a foot off the ground and flew the length of the runway. Next he taxied back and flew down the runway at a height of three or four feet. Then a flight that was higher still, about ten feet, during which he gently banked from side to side, feeling out the ailerons. After these preliminaries, as everything seemed in order, Ken flew a circuit around the field. Everyone cheered when he landed and taxied back. The little plane appeared to climb better and fly faster than similar small planes."

On May 16, 1936, CAA inspector R. A. Young observed test flights of the new plane and issued an Experimental License for the plane. This type of license did not allow for the carrying of passengers, so the plane could not be properly demonstrated to prospective buyers. On June 15, Taylor wrote the Bureau of Aeronautics in Washington, requesting permission to use the plane for demonstrations saying *"So far we have signed orders for over 73 airplanes and more coming in all the time, so we need this ship very much for a demonstrator."*

All this time the plane was being advertised in the Aviation press, and Taylor was accepting down payments for the new aircraft. As mentioned earlier, Bob Taylor believed his father had $10,000 in these advanced payments. But there was no real progress on either the financial or production fronts. The advance payments could not be used; they were supposed to be held in escrow. And D. Barr Peat's initial financing had run out, with no more in sight. Taylor was furiously scouring the country looking for new backers.

In June, Taylor happened upon George O'Brien, president of the Alliance, Ohio, Chamber of Commerce. O'Brien was what would have been termed in those days "a live wire". Even during that depression year he had been successful in attracting several new businesses to Alliance. When he and Taylor met, it was evident that they had mutual interests. Alliance needed new businesses to employ its hundreds of idle workers. Even more fortunate, it had a vacant but completely equipped aircraft plant!

In 1929, local investors had backed the Alliance Aircraft Corporation in its attempt to build and market the Argo biplane and the Hess Warrior engine. They had constructed a steel framed factory building, 60' wide and 300' long and equipped it with a number of machine tools. The corporation failed in 1930 and the plant had remained unoccupied since that time. It was an ideal situation for Taylor.

Once the negotiations started, they moved quickly. Taylor estimated it would take $50,000 to set up the new plant and begin production. The Chamber of Commerce agreed to raise this amount through local

The beginning of Taylorcraft production in Alliance ALLIANCE LIBRARY

stock subscriptions. So on July 8, 1936, trucks were dispatched to Butler to move the Taylorcraft factory, and the C. G. Taylor family, to Alliance, Ohio.

With a new plane now flying, and a factory at his disposal, Taylor expected to start production almost immediately. In fact he was quoted in the local newspaper as saying *"Production will begin within a week".* 150 new jobs would be created to handle the supposed $100,000 backlog of business already on hand.

Actual production, however, began with agonizing slowness. For one thing, they needed to acquire an Approved Type Certificate for the new plane before it could be sold in the commercial market. This meant a detailed stress analysis as well as a series of static tests on the structure. These went on through the Fall of 1936, but it was not until 12-17-36 that Group 2 Approval, (#2-529), was obtained. Production was almost non existent. The first plane made in Alliance, NC15757, S/N 26, was finished in October. Only four more were produced by the end of the year. By contrast, Piper made over 500 J-2 Cubs during the same period.

The first Taylorcraft actually sold went to dealer Nathan Kohn of Patterson, New Jersey. He had made a number of trips to the factory in the Fall of 1936, pushing for early delivery. He once remarked, *"I have orders for 31 ships right now and could get rid of 50 more right away."* Kohn's first plane, which had a bright green fuselage and silver wings, bore the CAA number NC15776.

By January of 1937, it would seem that the Taylorcraft organization was poised for success. They had a new plant set up, the public liked their new plane, and there was at least minimum financial backing from the Alliance businessmen. To handle sales, an experienced executive, Carl Wooten, was employed. His first task was to set up the company's display at the New York show. According to the local newspaper, the new Taylorcraft was *"The hit of the show and 32 new orders were received".*

C. G Taylor with the first Alliance-produced Taylorcraft. TAYLOR

The actual sales and production results were less promising. Although over 40 people were now employed, only three planes were made in January. Production remained a trickle through February and the first part of March. The initial financing had been used up, and the revenue from the company's few sales was not enough to carry the plant expenses.

But on March 16, 1937, an event took place in Bradford, Pennsylvania, which would project Taylorcraft into the top ranks of aircraft manufacturers. Word came to Alliance that the Cub plant had burned down! Piper, Taylor's main competitor, was out of business!

On March 20, the Alliance Review carried the following news headlines: "*NEED MORE FACTORY SPACE TO MEET RUSH ORDERS FOR PRODUCT. Taylor Cub Plant disaster leaves Taylorcraft alone in field*"

Sensing a golden opportunity, a group of local investors, working with New York financier F. L. Wehrham, immediately bought out the original investors and formed a new corporation. Now known as the Taylor-Young Airplane Corporation, William C. Young would be executive vice-president. Taylor was allowed to keep the title of president, but a number of experienced manufacturing men, headed by Stanley Vaughn, were brought in to speed production.

Certainly, more production was needed. In April, the company announced it had 257 unfilled orders. On May 13, the Alliance Review headlined, "*Factory nearing peak production. By June 12, twelve planes a week will be built, and 3000 will be produced in 1938. Employment has increased to 95, and a new paint shop will be built.*" At last, Taylorcraft was off and running.

On June 11, 1937, through the efforts of the engineering staff, Ray Carlson and Harold White, a full Approved Type Certificate, #643 was received. In keeping with the strategy of offering a more "luxurious plane than the Cub", many options were offered. The basic list price was $1495, but after options were added, the plane usually cost over $1600 when delivered.

The Sales Department found that more and more customers requested brakes and a tail wheel on their planes. Byron Shinn had responded to this need by designing an extremely simple and lightweight system for use on light aircraft. The Shinn System reversed the automotive practice of attaching the brake lining to the brake shoe. Instead, the lining was put on the drum, keeping the heat away from the lightweight tires. The actuating mechanism contained very few parts and no bearings to rust or bind.

Production and sales did jump dramatically. In June Nathan Kohn was again featured in the local press when he took delivery of the 100th Taylorcraft. He commented, "*The demand for the Taylor-Young plane in the east is astounding.*" By July the assembly line was putting out 50 planes per month, with 140 people employed.

The increase in production put severe strains on the financial structure of the new company. The sales volume was nearing $65,000 per month. The payroll was at least $10,000 and the bill from Continental for the engines would be near $20,000. Without sufficient working capital, the new owners found themselves continually short of cash. More investors were needed.

Taylorcraft employees in the Fall of 1937 — ALLIANCE LIBRARY

TAYLORCRAFT Model A 2-place **ATC 529**

Specifications. Span 36′ (11 m.) Length 22′ (6.7 m.). Height 6′ 8″ (2.1 m.). Wing area 155 ft.² (14.4 m.²). Power loading 26 lbs./h.p. (11.8 kg./h.p.) Wing loading 5.8 lbs./ft.² (29.3 kg./m.²). Weight empty 586 lbs. (265.8 kg.). Useful load 464 lbs. (210.5 kg.). Payload 226 lbs. (102.5 kg.). Gross weight 1050 lbs. (476.3 kg.). Baggage allowance 56 lbs. (25.4 kg.). Fuel 10 gals. (37.8 lit.). Oil 1 gal. (3.8 lit.).
Engine. Continental A-40-4 h.p. at 2575 r.p.m.
Performance. Maximum speed 91 m.p.h. (146.5 k.p.h.). Cruising speed 80 m.p.h. (129 k.p.h.). Landing speed 35 m.p.h. (46.3 k.p.h.). Service ceiling 14,000 ft. (4260 m.). Climb per min. 390 ft. (118.9 m.). Cruising range 230 mi. (371 km.).
Construction. Fuselage: welded steel tubing, truss type; fabric covered; side-by-side seating arrangement with dual wheel controls of the inset panel type. A large door is provided on the starboard side, hinged to swing forward. Wing: metal ribs; wood spars; fabric covered; bracing by a pair of steel tube streamlined V struts attaching to the bottom fuselage longeron aft of cabin door, and having vertical intermediate struts located at their centers. Tail group: welded steel tubing; fabric covered; braced by wires extending from the trailing edge of the stabilizer to the trailing edge of the fin, and to bottom fuselage longeron. Fixed divided type landing gear with shock cord absorption at the center. Welded steel tail skid swivelling through 120°.
Standard equipment. Sensenich metal-tipped wood propeller 69 inches in diameter, Silvertown 6.00 × 6 tires, plain oilless bearings in control system, wired for navigation lights and indirect lighting of the instrument panel, fire extinguisher, imitation leather upholstery, Plastecele windshield and cabin enclosure, removable dual wheel controls, Shinn Devices, Timken-bearing wheels.
Instruments. Aeromarine altimeter and airspeed indicator; Taylorcraft combination oil pressure, oil temperature gauge, tachometer.

Taylorcraft data AERO DIGEST

In the Fall, after threatening to move from Alliance, the company offered to sell additional stock to the local citizens. A "Fund Drive" was established by the Chamber of Commerce, with a goal of $115,000. Leading businessmen, fearing the loss of this large payroll, enthusiastically supported the drive, sometimes buying stock and giving it free to their employees. By December, it was clear that the needed funds had been pledged, and in January 1938, ownership again changed, this time to a group headed by A. A. Mulac, a prominent local manufacturing executive.

After slowing during the winter, sales again rebounded in the Spring of 1938. But it was evident that the days of the little 40 HP Model A Taylorcraft were numbered. Piper, Aeronca and Luscombe were all working on new models that would use the Continental A-50 engine. Taylor began working on the new Model B, which would offer the choice of 50 HP Lycoming, Continental or Franklin powerplants. When production ceased in the fall of 1938, six hundred and six "A"s had been produced. They had been delivered to every state of the Union and over a dozen foreign countries. Thousands of pilots had learned to fly in the gentle little A; they would always remember its honest, comfortable way of flying. Over the years, largely due to the lack of horsepower and marginal performance, most were scrapped. In 1997, only 32 remained on the FAA registration list.

Under various ownerships, Taylorcraft went on to build thousands of their attractive, good flying planes. During WW II they contrubuted over 2000 of their Tandem L-2's and 250 of their TG-6 gliders to the war effort. But sadly, like many light aircraft companies, they were shut down by bankruptcy in 1946. Charlie and Dorothy Ferris made a valiant attempt at a revival with their 100 HP F-19 in the 1980's, but finally sold out after building about 150 units. In all, Taylorcraft built almost 10,000 planes. Of that number approximately 3000 remain active today, (2001).

The Author and wife Marian with their F-19 in 1985 AUTHOR

Specifications
Taylorcraft "A"
ATC- 643, 6-11-37

TAYLORCRAFT A, 2 PCL-SM, TC 643

Engine Continental A-40-4 (See item 309 and NOTE 2)
Placard limits Maximum, except take-off
-- in. Hg., 2575 rpm (40 hp)
Take-off (one minute)
-- in. Hg., 2575 rpm (40 hp)
Propeller Maximum permissible diameter 75 in.
Placard speeds Level flight or climb 92 mph True Ind.
Glide or dive 125 mph True Ind.
Fuel capacity 10 gals. (-12)
Oil capacity 1 gal.
No. pass. 1 (+23)
Baggage 30 lbs. (+41)
Standard weight Landplane - 1050 lbs.
Seaplane - 1130 lbs.
C. G. limits Landplane (+16.0) and (+19.6)
Seaplane (+18.1) and (+20.4)
(with item 151)
(+17.0) and (+20.4)
(with item 351)
Spec. basis Type Certificate No. 643 (Aero. Bulletin 7A requirements)
Serial Nos. Landplane - 26 and up eligible per NOTE A
Seaplane (with item 151) - 362 only eligible (others eligible when fuselage conforms to Dwg. A-A110, change "O")
Seaplane (with item 351) - 489 only eligible (others eligible when fuselage conforms to Dwg. A-A165, change "C".)
(Manufacturer's affidavit required)

Equipment: (Datum is leading edge of wing)(*Means net increase)

Class I:
(a) Landplane
101. 6.00-6 wheels (Shinn)
102. Tail skid
103. Propeller - wood (Fixed or adj. pitch) 11 lbs. (-47)
(b) Seaplane; Item 103 PLUS
151. Edo D-1070 floats with water rudders (149 lbs.) } 109 lbs.*
152. Auxiliary seaplane fin }

Class II:
(a) Landplane
200. Misc. items as noted in approved weight and balance report.
201. Everel propeller I-38-A (single blade) 12 lbs.*
202. Radio receiver 9 lbs. (+38)
203. Battery and box 9 lbs. (-17)
(b) Seaplane: Items 200, 201, 202, 203 and 309b.

Class III:
(a) Landplane
301. Parachutes - one or two at 20 lbs. each
302. Tail wheel
(a) 6x2.00 steerable 3 lbs.*
(b) 6x2.00 steerable (Aircraft Associates) 3 lbs.*
(c) 6x2.00 (Goodyear commercial) free swiveling 1 lb.*
303. Auxiliary door
304. 6.00-6 wheels (Shinn) with brakes 5 lbs.*
305. Wheel pants 6 lbs.
306. (a) Cabin and carburetor heater (Wooster) 4 lbs.*
(b) Taylorcraft cabin heater A-166 2 lbs.
307. Navigation lights 3 lbs.
308. Miscellaneous extra instruments
309. Engines
(a) Continental A-40-3
Placard limits
Maximum, except take-off
— in. Hg., 2550 rpm (37 hp)
Take-off (one minute)
— in. Hg., 2550 rpm (37 hp)
(b) Continental A-40-5 (dual ignition) Approx. 11 lbs.*
Placard limits
Maximum, except take-off
— in. Hg., 2575 rpm (40 hp)
Take-off (one minute)
— in. Hg., 2575 rpm (40 hp)
310. Pressure cowling (Greider model A) 2 lbs.*
311. Skis Use actual weight change
(a) Federal SA-1
(b) Richards 1-B
(c) Air Transport 1220-480
(d) Alaska Airmotive Taylorcraft
(e) Jennings TY-2
(f) Federal SC-1
(g) Marston 1200
(b) Seaplane: Items 301, 303, 306, 307 and 308 PLUS
351. Edo 54-1140 float installation (incl. water rudders) 140 lbs.
(Approx. 100 lbs.* with item 152)

NOTE A. Each aircraft manufactured after April 3, 1941, must, prior to original certification, satisfactorily pass:
(a) An inspection for workmanship, materials and conformity before any covering, metal priming or final finish is applied. All woodwork may be varnished.
(b) A final inspection of the completed aircraft.
(c) A check of the flight characteristics.

NOTE 1. Eligible for export as follows subject to inspection for equipment specified in Chapter XII of Inspection Handbook: (April 1, 1942)
(a) Canada - Landplane (A-40-4 engine must be placarded for 39 hp at 2500 rpm)
Seaplane - maximum standard weight 1070 lbs. with item 151 and 1130 lbs. with item 351.
Skiplane - not eligible. However, structure complies with Canadian requirements for ski installation with ski height of 12 in. or less.
(b) All other countries except Australia and New Zealand.

NOTE 2. No aircraft of this model shall be eligible for original certification with single ignition engine after August 1, 1941.

Chapter Eleven

Porterfield CP-40 "Zephyr"

JUMPTNER

The prototype Porterfield Zephyr,NC16450, at Kansas City in 1936. It has the optional cabin enclosure.

The Porterfield Zephyr arrived on the aviation scene at the same time as the Taylorcraft and the J-2 Cub, yet it was a marketing failure, selling less than 20 units. Although priced competitively, ($1295 for the open model), its design was uninspired, actually a make-over of the American Eaglet introduced six years before. It did, however, pave the way for Porterfield's highly successful Collegiate series, which followed some months later.

E. E. "Ed" Porterfield was a native of Kansas City, MO. A successful Ford auto dealer, he became interested in aviation when in 1925 he traded barnstormer Blaine Tuxhorn a used Model T for flying lessons in a Curtiss Jenny. Sensing a business opportunity, he started his own flying school the same year, and shortly thereafter incorporated the American Eagle Company to build planes. Their first product, the Model A-1, was a typical OX-5 powered biplane. It received one of the first ATC's granted by the department of commerce, (#17), on November 11, 1927. Several hundred of this model were built in the next few years, their brightly engine-turned aluminum cowling an easily noticed trademark.

At first the company prospered, continuing to build their standard biplane design with various engine options. In 1930 they even ventured into the light aircraft field with their 30 HP Eaglet described in Chapter 1. As the depression deepened in 1931, sales dried up, finances remained strained, and a merger was arranged with the Lincoln Aircraft Corporation of Lincoln, NE. Mr. Porterfield was demoted from president of his own company to sales manager in the new organization, which immediately caused discord. By 1932, the company was facing bankruptcy proceedings, so Ed Porterfield left.

However, by 1934 he was back in the airplane business with a new monoplane, the Porterfield 35-70. This design was the work of Noel Hockaday who had previously worked for American Eagle. During the winter of 1932-33 the manual training students at Kansas City's Wyandotte high school built a small two place monoplane under the guidance of Hockaday. It flew reasonably well, first powered by the ill fated Poyer engine, then by an Aeromarine 3-cylinder radial of 40 HP.

American Eagle biplane MALOY

Ed Porterfield was looking for fresh designs to launch his new company, so he struck a deal with Hockaday. In exchange for 200 shares of stock in the Porterfield Aircraft Corporation, (and a job), Hockaday turned over the plans and prototype airplane. Slightly redesigned to take a LeBlond 5-cylinder radial engine of 70 HP, the plane was christened the Flyabout 35-70.

The new plane was a modest success, due largely to Porterfield's excellent reputation in the aircraft field. After its introduction at the 1935 Detroit show, enough orders were received to set up serial production. The plane appealed to those pilots wanting more performance than offered by the Cub or Aeronca, and were willing to pay for the modest additional cost. It had gentle flying characteristics, not too challenging for those who were accustomed to the tiny Cubs, Eaglets, or the like. Over 200 of this model were built before it was discontinued in 1937.

Being well aware of market conditions in the aircraft industry, Porterfield knew that by 1936 the Cubs and similar lightplanes were gaining a large share of the sales. In order to capture the beginning fliers who bought such planes, he decided to again enter the low-horsepower field.

Hockaday had done much of the design work on the ultra-light Eaglet when he was with American Eagle. In 1936, knowing of this experience, Porterfield asked him to lay out a similar plane to use the now-popular Continental A-40.

Porterfield 35-70 *Juptner*

Open Model Zephyr JUPTNER

Not surprisingly, the new plane greatly resembled the Eaglet. Its wingspan and length were the same, as was the two place tandem seating arrangement. The rudder, horizontal stabilizer and elevators are Eaglet-type, with cambered airfoils built up of small steel tubing. The wings were also the same, wood spars and ribs, M-series airfoil, but the Zephyr had a fuselage-wide center section. The landing gear was of the common split-axle type, using the same wheels and tires as the Eaglet. The fuselage seemed to be pure 35-70 which may account for the 70-pound weight increase. The plane was to be offered in either a closed cabin or open configuration.

The only original design involved the firewall-forward portion; the installation of the Continental A-40-4 engine. Just as this work got underway, Hockaday left for the west coast, leaving Howard Barlow to carry on in his place. Although advertised to the trade in September of 1936, the plane did not receive its Group II approval #2-530 until 2-16-37.

PORTERFIELD Zephyr **2-place** **Approval 2-530**

Manufactured under Group 2 approval.
Specifications. Span 34 feet 8 inches (10.6 meters). Length 21 feet 7.5 inches (6.6 meters). Height 6 feet 11.25 inches (2.1 meters). Wing area 169 square feet (15.7 square meters). Power loading 26 pounds per horsepower (11.8 kilograms per horsepower). Wing loading 6.17 pounds per square foot (30.1 kilograms per square meter). Weight empty 578 pounds (263 kilograms). Useful load 462 pounds (209 kilograms). Payload 180 pounds (81.6 kilograms). Gross weight 1040 pounds (472 kilograms). Baggage allowance 10 pounds (4.5 kilograms). Fuel 9 gallons (34.1 liters). Oil 1 gallon (3.78 liters).
Engine. Continental A40-4, 40 h.p. at 2575 r.p.m.
Performance. Maximum speed 85 m.p.h. (136.7 km./hr.). Cruising speed 75 m.p.h. (120.5 km./hr.). Landing speed 35 m.p.h. (56.2 km./hr.) Service ceiling 9000 ft. (2740 m.). Climb per min. 425 ft. (130 m.). Cruising range 225 mi. (362 km.).
Construction. Fuselage: welded chrome molybdenum steel tubing, Warren truss type; tandem seating arrangement with removable dual controls; baggage compartment accessible in flight; fuselage fabric covered. Wing: spruce spars, built-up ribs, fabric covered; strut braced by streamlined steel tubes on either side of fuselage. Tail group: chrome molybdenum steel tubing, welded; fabric covered; stabilizer adjustable in flight from cockpit; wire braced to fuselage by Macwhyte tie rods. Fixed split type landing gear.
Standard equipment. Fahlin or Sensenich wood propeller, Goodyear tires, Goodyear wheels, Goodyear rubber disk shock absorbers, plain engine and plane control bearings, Pyrene fire extinguisher, first aid equipment, Pyralin windshield and cabin enclosure.
Instruments. Zenith aneroid altimeter; float-type fuel gauge; U. S. oil pressure and oil temperature gauges; Jones centrifugal type tachometer; connection for pitot static airspeed indicator.

Porterfield Zephyr Specifications AERO DIGEST

The CABIN ZEPHYR---$1,355.00

UNEQUALED FOR DUAL INSTRUCTION OR BUILDING UP YOUR FLYING TIME!

CONVENIENT TERMS

ONLY $432.00 Down Payment is required on the OPEN ZEPHYR, or $452.00 Down Payment on the CABIN ZEPHYR plane, with twelve monthly payments of $83.60 on the OPEN plane and $90.30 on the CABIN plane, including all finance charges. These convenient terms are available to those whose credit standing is A-1. For average or fair credit rating, the minimum down payment is 40 per cent, and the monthly payments are correspondingly less.

The OPEN ZEPHYR
$1,295.00

The PORTERFIELD ZEPHYR

Is by far the lowest priced airplane on the market today, and is manufactured by a reputable factory which has an excellent record for manufacturing aircraft. More than 1,000 airplanes have been manufactured by PORTERFIELD. This record should satisfy any prospective airplane purchaser on the PORTERFIELD REPUTATION.

Satisfactory test flights have been completed on this airplane and the stress analysis is now being carefully checked by the Bureau of Air Commerce, from which we receive satisfactory reports every week. Several orders are also being received each week, all of which will be delivered in the rotation in which they are received, after Government Approval is obtained. The PORTERFIELD factory now has three current Government Approved Type Certificates. Porterfield's former company obtained ten Certificates. This information is given to assure purchasers that we know what is necessary to obtain Government Approved Type Certificates.

PORTERFIELD
AIRCRAFT CORPORATION
KANSAS CITY, MISSOURI

Porterfield Zephyr ad in 1936. They almost offer an apology for the lack of an ATC.

AUTHOR

Lebrecht's restored Zephyr LEBRECHT

It was originally offered as either a closed or open model, but records indicate none of the open models were sold. In fact the prototype NC16450, pictured on page 99 of this chapter, was returned to the factory and became a closed model with the same number. (See photo on page 97).

The Zephyr was not a success. By the summer of 1937, Continental was offering prototype samples of their A-50 engine to most light aircraft manufacturers. As we will see in chapter 13, Aeronca was the first to have a 50 HP model certified; their KCA in April of 1938. Clearly, the days of the A-40 powered planes were numbered.

Approximately 20 were sold through 1937-38, certainly not enough to be considered profitable. Some reports mention the factory test pilots did not like the plane. Perhaps this information filtered on to the customers and affected sales. However, Chuck Lebrecht owns a well-restored Zephyr, NC18743, and reports it flies well.

As mentioned in the beginning of this chapter, the Zephyr led Porterfield into the low-horsepower field, and ultimately into the highly profitable CPT market. Based somewhat on the Zephyr design, the Porterfield Collegian CP-50, (Later CP-65) appeared in late 1939. Over 400 of this well-liked model were built before WWII halted production.

Specifications
Porterfield CP-40
TC 2-530, 2-16-37

MODELS: Porterfield CP-40 and CP-40A, 2 PCLM

T.C. NUMBER: 2-530

I - SPECIFICATIONS PERTINENT TO ALL MODELS:

Propeller	Maximum permissible diameter 80 inches
Airspeed limits	Level flight or climb - 83 mph True Ind.
	Glide or dive - 112 mph True Ind.
Fuel capacity	9 gallons (-10)
Oil capacity	1 gallon (-35)
No. passengers	1 (+8) or (+37)
Baggage	10 lbs. (+63)
Standard weight	1040 lbs.
C.G. limits	(+13.5) and (+21.1)
Specification basis	Aero. Bulletin 7A, Section 3
Serial Numbers	250 and up manufactured prior to 7/1/40 eligible. Approval expired as of that date.

EQUIPMENT: (Datum is wing leading edge) (* Means net increase)

Class I:			
101.	6.00-6 wheels (Shinn 6C2) with 2-ply tires	21 lbs.	(+3)
102.	Tail skid		
103.	Cabin enclosure	20 lbs.	(+28)
104.	Propeller - wood	9 lbs.	
Class II:			
200.	Miscellaneous items as noted in approved weight and balance report.		
Class III:			
301.	6.00-6 wheels (Shinn 6C2B) with brakes and 2-ply tires	6 lbs.*	(+3)
302.	16x7-3 wheels (Goodyear 3LMBM) and tires	2 lbs.*	(+3)
303.	16x7-3 wheels (Goodyear 3LMBM) and tires and brakes	13 lbs.*	(+3)
304.	Carburetor air heater (Drawing B-804)	2 lbs.	(-36)
305.	Emergency exit (large window on left side)	2 lbs.	(+28)
306.	Navigation lights	2 lbs.	
307.	Miscellaneous extra instruments	4 lbs.	

NOTE 1. Following placard is required: "When flying solo, pilot must fly from front seat", unless an actual C.G. check is made substantiating the rear seat position.

NOTE 2. Eligible for export as follows, subject to inspection for equipment specified in Chapter XII of Inspection Handbook (10/12/38):

(a) Canada
- Landplane - Provided item 305 is installed and the following engine limits use for Model CP-40: For all operations, 2500 rpm (39 hp).
- Skiplane - not eligible.

(b) All other countries except Australia and Great Britain.

II - MODEL CP-40:

Engine	Continental A-40-4
Engine limits	For all operations - 2575 rpm (40 hp)

III - MODEL CP-40A:

Engine	A-40-5
Engine limits	For all operations - 2575 rpm (40 hp)

Chapter Twelve

Welch OW5M

This prototype Welch monoplane sports a rather garish paint scheme. The motor is a very early A-40.

DRINA WELCH-ABEL

Orin Moore Welch was born in Ohio in 1906, at the time the Wright brothers were making their first flights near Dayton. Like many young men of that era, he was interested in all new things mechanical; at the age of ten, he wired the Welch home for electric lights, and began to experiment with some of the first crystal radio sending and receiving sets.

But in 1920, when his father gave him five dollars for an airplane ride, he was bitten by the aviation bug, and from then on Orin Welch would make aviation his life. Learning that Mr. L. H. Scott was going to locate a flying and repair business in their home town of Marietta, OH. Orin immediately left school, (he had finished the eighth grade), and went to work in this new enterprise. Scott was a good teacher, and by 1923 Orin had become an expert mechanic and had soloed in Scott's Curtiss Canuck.

Shortly after this solo flight, Orin's father paid $1800 for an OX-5 powered Standard J-l, and this seventeen year-old youth started a barnstorming operation. In 1927, the family moved to Anderson, Indiana, where They took over the operation of the Anderson Aircraft Company. In addition to a large, city owned airport, they had the use of a spacious hangar and several shops. It was here that Orin, still only twenty-one years old, would begin his aircraft manufacturing activities.

His first designs were traditional three-place biplanes, powered by war-surplus OX-5 engines. While they performed well, they offered no special improvement over the many other similar designs made by Waco, TravelAir and others. Only a handful were built.

But just as the Anderson operations were beginning to prosper, tragedy struck. In November of 1929, the main hangar including almost all the Welch planes and equipment, was destroyed in a spectacular fire. They were out of business. For the next two years Oren was forced to take whatever work he could find, in or out of aviation, to survive.

While engaged as a company pilot on the west coast, Orin designed a small monoplane, intended to offer economical flying in the now cash-poor depression economy. His layout looked much like an Aeronca C-3, and had many of the same characteristics.

IF YOU CAN DRIVE A CAR - - YOU CAN FLY THE NEW FALCON

THE NEW FALCON 2-PLACE CABIN COUPE

★ ★ ★

ACE AIR-CRAFT

"THE ACE IN AIRPLANE EXCELLENCE"

UNFALTERING PERFORMANCE . . . . BACKED BY PROOF AND TESTS

● Recently, at the Municipal Airport at Indianapolis, a New Falcon, 2 place cabin coupe was thoroughly subjected to every flying test. A 45-mile gale and driving rain kept most air-wise pilots quizzically scanning the troposphere, while the New Falcon made more than thirty record trips, verifying basic construction specifications—extraordinary strength of wing and fuselage—an unfailing motor with abundance of power—simplicity of control—and proving positive operation at all speeds, due primarily to ample control surfaces and perfect balance throughout. This exhaustive testing provides advance experience which assures outstanding VALUE of the New Falcon in safety and performance.

* * * * *

Spring stands expectant, just beyond the distant hills; now is the time to plan for the days that soon will come. Investigate the New Falcon! It is economical, easy to handle, and exceedingly comfortable; a ship you can be sure of and proud of always.

$995
(FLY-AWAY FACTORY)

PLACE YOUR ORDER EARLY—DELIVERIES CAN BE MADE IN ROTATION ONLY!

ANYONE CAN QUICKLY LEARN TO OPERATE THE NEW FALCON SAFELY!
2 to 5 hours of free instruction from your dealer, or our own pilots will enable you to master this simple, new modern means of transportation. If you can drive a car you can fly the new Falcon.

The New Falcon is a two-place Cabin Coupe with a 42 horsepower two cylinder air-cooled engine of our own make, gas consumption 3 gallons per hour, low speed 28 miles, top speed 85 miles, cruising speed 75 miles. Wing spread 36 feet, chord 50 inches, length 20 feet 6 inches, height 5 feet 8 inches. Steel hangar 10 feet by 24 feet manufactured for the Falcon, $150.00.

Optional Installation—Continental A-40, 4 cylinder motor $200 additional.

AIR-CRAFT CORPORATION OF AMERICA

General Offices: Majestic Bldg.
INDIANAPOLIS, IND.
● ● ●
Factories: Portland, Ind.

DEALERS! *You Can Sell this Ship*—
Attractive proposition with possibilities unlimited. We will outline plans which will bring you real profits! Exclusive sales franchise guaranteed. Write or wire today.

ACE AIR-CRAFT

USE THIS COUPON 3-32
AIR-CRAFT CORP. OF AMERICA
8th Floor—Majestic Building
Indianapolis, Indiana
Send details, I am interested as, ☐ Dealer ☐ Private Buyer

NAME . . . .
ADDRESS . . . . STATE . . . .

In 1931, Orin Welch designed the "Falcon", a light two-place monoplane, and obtained financial backing from a group of Indianapolis investors. This ad appeared in the March 1932 Popular Aviation.

POPULAR AVIATION

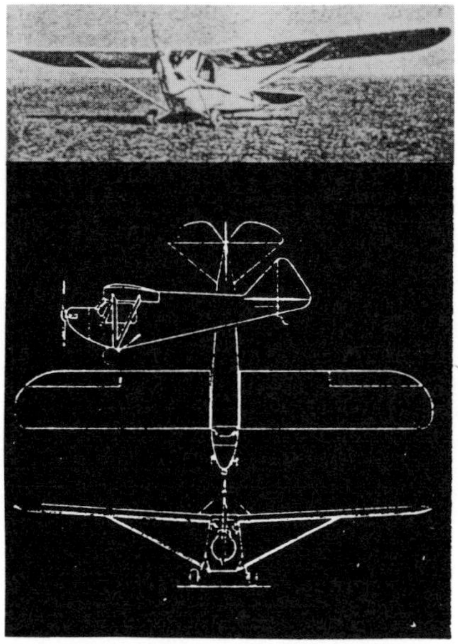

WELCH Models OW5M and OW7M

Orin Welch Aviation School, Anderson, Indiana

• President and Chief Engineer: Orin Welch. Sales Manager: John Welch.

One- or two-place open or closed monoplane. (Models OW5M and OW7M respectively; data on OW7M indicated in brackets.) Continental A-40 engine, 37 horsepower (Welch 0-2 engine, 45 horsepower). Approval number 2-474. Also available with Szekely engine, 45 horsepower; Salmson AD-9 engine, or Aeronca 113A engines. Span 34 feet 4 inches. Length overall 21 feet. Height overall 5 feet 8 inches. Wing area 136 square feet. 26.5 pounds per horsepower. Chord 54 inches. Wing loading 6.6 pounds per square foot. Empty weight 540 pounds. Useful load 394 pounds. Payload 170 pounds. Gross weight 950 pounds. Fuel capacity 8 gallons. Oil capacity 1 gallon. Maximum speed 93 miles per hour. Cruising speed 85 miles per hour. Landing speed 29 miles per hour. Service ceiling 11,000 feet. Rate of climb 450 ft./min. Cruising range 200 miles.

Fuselage: steel tubing, with fabric covering; three longerons, one longeron on the bottom extending back from cabin; seating arrangement in two place models are side-by-side; single control stick; manufactured in both open and closed models. Wing: fabric covered; spars are built up from two-play spruce web with spruce cap strips; ribs are of basswood; dural drag struts are used with drag wires; large diameter chrome molybdenum strut wing bracing converging at a large fitting on the fuselage, which is also used to secure the top axle fitting. Tail group: chrome molybdenum steel tubing; wire bracing, adjustable by placing a bolt in one of a series of holes in a plate fitting at the front of stabilizer. Fixed type landing gear equipped with 16 × 7.3 airwheels, no brakes. Controls are operated by cable on plain bearing.

Standard equipment includes Flottorp propeller and an ignition switch.

These specifications for the Welch OW5-M and OW7-M were published in 1935. Customers were given a wide choice of engines. AERO DIGEST

In 1931, Oren returned to Indiana and was able to interest a group of Indianapolis financiers in manufacturing the little two place ship. Space was leased in Portland, Indiana, and initial production work was started. It was to be a very small, inexpensive machine; with a target retail price of $995. First flown with a two-cylinder engine made by Welch, it was later powered with the new Continental A-40. A national advertising campaign was begun, but no sales resulted, so the company was dissolved. The ownership of the plane reverted to Orin. He removed the Continental engine, installed a 6-cylinder Anzani, and sold the plane.

Orin had always remained interested in light aircraft design. He wanted to manufacture planes, which meant an Approved Type Certificate was essential. With this in mind, he set up a shop and aviation school back in Anderson. There he built a second lightplane, similar to the Falcon, but with a number of improvements. The structure was strengthened, and the cabin space widened for more comfort. Fairing strips improved the streamlining. The Continental A-40, originally used in the Falcon, was installed in the new plane. When finished in white dope with red stripes, it had a cute, "perky" appearance. This plane, NC13500, S/N 108, is shown at the start of this chapter. It was used to obtain Approved Type Certificate #637 which was not issued until 5-17-37.

By the time this approval was received, 1937, time had almost run out for small 40 HP lightplanes. Continental, Lycoming and Franklin were soon to introduce their 50-65 HP models, which offered the lightplane flier greatly increased performance. Nevertheless, Welch continued to manufacture small numbers of these planes, perhaps thirty in all. (In December of 1939, the Author, then a college student, received a 4-page color brochure from Welch showing the OW-5M and OW-7M. The price was still $995).

Construction of the Welch monoplane was traditional in most respects. One unique feature was the control wheel hanging from the roof of the cabin. For some first-time users, this proved to be confusing.

Although the Welch family had eked out a bare living from their Anderson, Indiana operations, they were never able to begin any real volume of aircraft production. Oren assigned serial numbers to his planes in exact rotation, the first a biplane made in 1927 being S/N 100. By the spring of 1936, he had built only 12 planes in all, the last was S/N 112, an OW6M with an Aeronca E113A engine installed. The facilities at Anderson were not adequate for any sizeable production, and the city was too small to support a flying school. A move was in order.

The Welch factory in South Bend WELCH BROCHURE

The family was also able to occupy a vacant apartment on the second floor of the old factory, saving additional rent. As production increased the entire family began working in the enterprise. Orin's father did most of the woodworking while his mother sewed the fabric covers. Sister Violet handled the office and in addition did some of the flying. Even little sister Drina, then twelve, helped by wet sanding the doped surface to obtain the desired shiny finish.

From the factory records, it is evident that Orin Welch was more interested in designing planes, tires and motors, than in the serial production needed for profitable operations. One customer remembers the difficulty in getting a new Welch delivered in 1940, after the plant had moved to Wilkes-Barre, Pennsylvania.

"I was 19 in the summer of 1940 when Dad and I bought a new Welch OW-5. Orin kept putting back the delivery date on our plane and my dad kept pushing him until finally he said it would be ready in about ten days. A friend and I left for Wilkes-Barre about a week or so later to ferry the plane back to Dallas. We stayed about a month until the plane was ready! Every day we would go out to the factory to

In the summer of 1936, the aviation commission in charge of South Bend's Bendix Municipal airport offered the Welch organization space and other inducements if they would locate in their city. It was a welcome offer, quickly accepted. The local newspaper announced:

"Oren Welch, president of the Welch Aircraft Company, stated this morning that production of a low-cost airplane will get under way as soon as the manufacturing equipment can be set up and the organization completed. The company intends to hire additional help from among local men as soon as production requires a larger staff.

"The firm has manufactured the Welch sport monoplane for the last eight years in Anderson. The plane, a specialized product of the firm, is powered by a 45-horsepower engine and sells for about $1000 in competition with other low priced ships."

Oren had originally planned to occupy space offered at the Municipal airport, but later decided to move into part of a vacant watch factory at 1720 Mishawaka Avenue in South Bend. Here they had adequate factory space, but did have to truck the dismantled planes to the Glen Moon airport for assembly. For the first time a semblance of volume production was begun; some thirty planes were built between early 1937 and December of 1939.

This Welch appeared at a mid-west air meet in the 1980's. The A-40 has been replaced by a Lycoming 0-145 AUTHOR *watch the progress, and I was quite impressed to learn that the Welch even had its tires made right there.*

"I well remember the friendly, helpful people we met on that trip, and especially the folks at Phil's Diner in Wilkes-Barre. When we ran out of money for a few days, they let us eat on the cuff, and one night hosted a night on the town."

It was true that the Welch was probably one of the cheapest new planes available, selling for between $800-$1000. However, a customer evidently had to be very patient regarding delivery.

Orin Welch not only wanted to build planes, for some years he had attempted to make an aircraft engine as well. In 1929 he had designed a small two-cylinder model, the Welch O-2, which produced 45 HP at 2500 RPM. It was light, weighing only 115 pounds dry, somewhat less than the Continental A-40. It resembled the Aeronca E-113 and the Wright-Morehouse in many respects. Whatever its virtues or faults, it did not catch on in the aircraft trade. FAA records indicate that only six Welch planes were equipped with this engine.

The flying quality of the Welch monoplane remains somewhat of a mystery. Some old-timers claim it flew as well as an Aeronca C-3 but was slower. Several mention some unpleasant stalling characteristics. The plane did have an unusual amount of dihedral which would have made it quite stable in rough air. Its general layout and dimensions are similar to most lightplanes except for the wing chord. It was quite narrow, giving the plane only 138 square feet of lifting area.

The pricing policy of the Welch company must have led to continual financial difficulty. The list price varied from $795 to $995. Discounts were given to dealers, 5% for ordering two planes, 10% for three, etc. The average revenue from each sale could not have been much over $900. Subtracting the $400 cost of an A-40 engine, leaves only $500 to pay for instruments, tires, propeller, dope & fabric, steel tubing, and all other material and labor. It would appear they came close to losing money on each sale. True, the family furnished most of the labor, but it also should be computed into the direct cost. During the same period, the Cub and Aeronca planes were selling for more, and they had the advantage of volume production. Certainly the Welch Aircraft Company was no bonanza.

In May of 1940 the company made its final move, to Wilkes-Barre, Pennsylvania. Only Orin and his sister Violet went; the rest of the family remained in Indiana. It was the beginning of the end for Welch Aircraft. The factory was set up in a large hangar at the airport and limited production was started. The first plane, christened the "Wyoming Valley", S/N 1006 NC28809, was exhibited in the town square amid speeches by various local dignitaries. But with Piper, Taylorcraft and Aeronca now selling their 65HP models, few customers were interested in the Welch using the obsolete A-40. Only six planes were made before Orin sold out to the Hartman Aircraft Company.

The new company managed to assemble one airplane, S/N1011 NC33500, before it too ceased production. A half-dozen Welch OW5M models still remain in the hands of antique plane collectors.

WELCH O-2

Welch Aircraft Motors, Anderson, Indiana

• President, General Manager and Chief Engineer: Orin Welch. Vice-president: Gomer Welch. Sales Manager: John Welch.

Two-cylinder, opposed, direct-drive, air-cooled. Rated by the manufacturer at 45 horsepower at 2500 revolutions per minute and 41 horsepower at 2300 revolutions per minute, using 72 octane fuel. Dry sump, full pressure feed lubrication. Compression ratio 5:2. Bore 4.5 inches. Stroke 4.25 inches. Displacement 135 cubic inches. Length overall 24.5 inches. Height overall 25.5 inches. Width overall 37.75 inches. Bare weight 115 pounds. Weight in pounds per horsepower at sea level 2.4. Fuel consumption at rated horsepower .64 pounds per brake horsepower per hour. Oil consumption at rated horsepower .02 pound per brake horsepower per hour.

Construction: Crankshaft nickel steel. Crankcase aluminum alloy two piece type, mounting is by four nickel steel studs placed in the bottom of the upper crankcase. Pistons aluminum alloy trunk type with two Perfect Circle compression and one Perfect Circle oil ring to each cylinder. Cylinders are of single alloy casting including barrel, head and rocker box; bronze guides and seats are screwed and shrunk into place and a steel sleeve is shrunk and pressed into place for piston travel. New Departure bearings. Wilcox-Rich valves, one intake and one exhaust to each cylinder.

Welch engine specifications — Aero Digest

Welch OW5M
ATC # 637
5-17-37

HARTMANN AIRCRAFT CORPORATION

MODELS: Hartman OW5M (Formerly Welch), 2 PCLM

T.C. NUMBER: 637

Engine	Continental A-40-4 (See NOTE 1)
Fuel	73 minimum octane aviation gasoline
Engine limits	For all operations, 2575 rpm (40 hp)
Airspeed limits	Level flight or climb - 75 mph True Ind. Glide or dive - 90 mph True Ind.
Propeller limits	Maximum permissible diameter 77 in.
C.G. range	(+17.3) to (+19.5)
Datum	Leading edge of wing
Maximum weight	950 lbs.
No. seats	2
Baggage	9 lbs. (behind seat)
Fuel capacity	9 gallons (two 4-1/2 gallon tanks in wing L.E.)
Oil capacity	1 gallon (engine sump)
Serial Nos. eligible	115 and up (See NOTE A)
Required equipment	Items 101, 102, 103
Certification basis	Type Certificate No. 637 (Aero. Bulletin 7A requirements)
Production basis	None (See NOTE A)

Eligible for export to all countries except Great Britain, Canada, Australia and New Zealand.

EQUIPMENT:

(Items included under "Required equipment" may not be removed unless replaced by approved equivalent items. The effect upon balance of all equipment changes must be computed and the aircraft operation record revised accordingly. Values in inches shown in parenthesis after each item represent horizontal arms to the C.G. of the item measured minus (-) ahead and plus (+) to the rear of the datum. A plus (+) or minus (-) sign preceding the weight of an optional item indicates the net weight change between that item and the equivalent required item.)

Propellers and Propeller Accessories

103.	Propeller - wood (fixed or adj. pitch)	

Landing Gear and Floats

101.	16x7-3 wheels (Goodyear 3NBA)	12 lbs.
102.	6x2.00 tail wheel	
303.	Hartman Model "C" (formerly Welch) cushion wheel	-1 lb.

Miscellaneous (not listed above)

200.	Miscellaneous items as noted in approved weight and balance report.	
301.	Parachutes (1 or 2)	20 lbs. each
302.	Streamlined lift struts	-12 lbs. (approximately)
303.	Skis (a) Heath 725	Use actual weight change

NOTE A. Prior to original certification, each aircraft must satisfactorily pass:

(a) An inspection for workmanship, materials and conformity before any covering, metal priming or final finish is applied. All woodwork may be varnished.

(b) A final inspection of the completed aircraft.

(c) Check of the flight characteristics.

NOTE 1. No aircraft of this model shall be eligible for original certification with single ignition engines after 8/1/41. In addition, no aircraft of these models shall be eligible for recertification with single ignition engines unless such aircraft were either previously certificated with single ignition engines or were originally certificated prior to 8/1/41.

Chapter Thirteen

Aeronca KC

One of the first Aeronca KC models, S/N KC-13, NC19775 JUPTNER

Aeronca's KC model was a perfect match-up for the Continental A-40. Its sleek lines, and well streamlined engine cowl, made it one of the most sprightly looking planes of its day. It brought the company into the "modern" aircraft era. And, as various higher horsepower engines became available, it evolved, through the KCA, into Aeronca's highly successful Chief series.

As mentioned in Chapter 1, the Aeronautical Corporation of America, (Aeronca), was formed by a group of Cincinnati, OH businessmen to exploit the 1928 "Lindbergh Craze" in aviation. At the same time, a group headed by Jean Roche had been attempting to market a diminutive monoplane in nearby Dayton. They joined forces and introduced their single-place Model C-2 in the early months of 1930. It was perhaps the first true lightplane certificated by the Department of Commerce. Using a 26 HP derivative of the Wright-Morehouse two cylinder engine, it offered a single place, later a two place, minimum sort of plane. And, unique among aircraft manufacturers, Aeronca manufactured their own engine, first the E-107 of 26 HP, and later a larger model, the 36 HP E-113.

Because of their low price, and reasonable performance, over a thousand C-2's and C-3's were built between 1930 and 1937. They brought flying into the reach of many during those depression years with their low operating and maintenance costs. The rather odd-looking wire braced monoplanes had become a common sight in the skies above America's airports.

But by 1937, the Aeronca executives could see that their primitive little plane was being outperformed, and outsold, by other competitive lightplanes. Thousands of Cubs were now in the air; selling for hundreds less than the Aeronca C-3. C. G. Taylor had started to build the Taylorcraft Model A in Alliance; it was better looking, and more comfortable to fly than most other lightplanes. If Aeronca was to continue its success, a new plane would have to be offered.

An early Model C-2 Aeronca JUPTNER

Their answer was a little strut-braced high wing monoplane, the "K", which was still powered by their E-113 two cylinder motor. Like Taylorcraft, it attempted to imitate larger, more expensive planes, with a well appointed cabin and wheel controls. The "K" was similar in most respects to its competitors, Piper and Taylor. Only one feature, the wooden wing structure, could be considered outdated. Both Piper and Taylor used the much superior aluminum ribs in their wing construction.

The Model "K" received ATC #634 on 4-30-37 and was an immediate success. With a price of $1745, it was slightly more expensive than the Cub and Taylorcraft, but sales were brisk for the first few months after its introduction, partly due to a loyal dealer organization.

However, as the fall of 1937 approached, all lightplane manufacturers were awaiting the announcement of new, more powerful engines from Continental, Lycoming and Franklin. This put the Aeronca people in a quandary. Since the start of their production seven years ago, they had relied on their own two-cylinder engine to power their planes. They had managed to increase it's output to an honest 45 HP, but that was the limit. If they needed more power, a new engine would have to be found.

There were other factors in play. While the Aeronca engine was an adequate powerplant, it did have it's troubles. Crankshaft occasionally cracked, requiring an expensive replacement. Since Aeronca bought all the parts for the engine, and only did the assembly, the cost was high. Also, more and more people were openly critical of a "two cylinder flivver airplane". Their only obvious choice was to purchase a complete engine from outside the company.

At that time, even though other engines were about to be introduced, the Continental A-40 was the only motor in wide use by the industry. It had been adopted by the two main players in the field, Cub and Taylorcraft. Even though it had slightly less power than their E-113 engine, the Aeronca engineers decided to install the A-40 in their new model "K".

The result was the Aeronca "KC" which received ATC #655 on 10-2-37. The first KC, S/N KC-1 NC19747, was completed on 8-25-37. Basically, it was a stock K model with the A-40 installed in a neatly formed front cowling. The only other obvious change was a new split-axle landing gear which increased the tread from 64 inches to 70 inches.

The structure of the fuselage consisted of a welded triangular frame of chrome molybdenum steel tubing, but the triangular structure was faired to a streamline shape by wood framing. Wings were built up with solid spruce spars and truss-type spruce ribs, braced by aluminum alloy V-struts. The tail group was also welded chrome-moly steel tubing, braced by streamline wires. The port side elevator was provided with a trim tab controlled from the cockpit. The entire plane, except for the nose section, was fabric covered.

Aeronca C-3 "Razorback" JUPTNER

AERONCA KC **2-place** **ATC 655**

Specifications. Span 36′ (11 m.). Length 20′7″ (6.3 m.). Height 6′7″ (2 m.). Wing area 146.35 ft.^{2} (13.6 m.^{2}). Power loading 26.5 lbs./h.p. (12.1 kg./h.p.). Wing loading 7.9 lbs./ft.^{2} (38.8 kg./m.^{2}). Empty weight 610 lbs. (277.6 kg.). Useful load 450 lbs. (204.1 kg.). Payload 214 lbs. (97.1 kg.). Gross weight 1080 lbs. (481 kg.). Baggage 44 lbs. (20 kg.). Fuel 10 gals. (37.8 lit.). Oil 1 gal. (3.8 lit.).

Engine. Continental A-40-4, 40 h.p. at 2575 r.p.m.; also Continental A-40-5 (dual ignition).

Performance. Maximum speed 93 m.p.h. (149.7 k.p.h.). Cruising speed 85 m.p.h. (136.8 k.p.h.). Landing speed 35 m.p.h. (56.3 k.p.h.). Service ceiling 12,000 ft. (609.6 m.). Rate of climb 450 ft./min. (2.3 m./sec.). Cruising range 250 mi. (402.3 km.).

Construction. Fuselage: welded chrome molybdenum steel tubing; wood fairing, fabric covering; detachable engine mount; four-longeron cabin, three to rear; fin integral with fuselage; side-by-side seating with dual inset panel type wheel controls; one-piece windows, adjustable to any position; door on starboard side; baggage compartment behind seat. Wing: solid spruce spars, single drag truss type; spruce truss ribs with fibre gussets; fabric covering; aluminum leading edge; Clark Y airfoil section; Frieze metal ailerons; bracing by aluminum alloy Vee streamlined struts to fuselage. Tail group: welded steel tubing; fabric covering; trimming tab on left elevator; bracing by streamlined wires. Tripod oleo type landing gear. Swiveling tail wheel.

Standard equipment. Sensenich wood propeller, Firestone tires, Shinn wheels, oleo shock struts, Olite bearings, Shakespeare throttle controls, Pyralin windshield and cabin enclosure, Dietz fire extinguisher, engine primer pump, first aid equipment, Blocksom Paratex cushions.

Instruments. Aeromarine altimeter, airspeed indicator; Stewart-Warner tachometer; oil pressure, oil temperature and fuel gauges.

Other data in November, 1937, AERO DIGEST.

Aeronca "K" Announcement ad and published specs

AERO DIGEST

An Aeronca "K" at a mid-west fly-in. This plane has had a Lycoming engine installed to replace the E-113 Aeronca built two-cylinder model. AUTHOR

The cabin was completely upholstered, adding to the appearance and the comfort of the passengers. Interior appointments followed the current automobile practice, much like the Taylorcraft. Large one piece windows, adjustable to any position, could be raised and lowered in automobile-type felt window channels. Ventilation could be controlled by means of these windows. Entrance was through a wide, almost square, door on the right side, and an optional left-hand door was offered.

The seat was a canvas sling type equipped with two paratex seat cushions. The cushions could be removed if the occupants elected to wear parachutes. Width and length of the cabin offered average shoulder and leg room. A small baggage compartment with a twenty-pound capacity was located behind the seat. Visibility was enhanced by a large sloping windshield, an amber-tinted overhead window, and streamlined rear windows.

The "K" had a modern-looking instrument grouping in the center of an oval shaped panel, which included a tachometer, oil pressure and oil temperature. To the left was an airspeed indicator and to the right a non-sensitive altimeter. The throttle was in the lower center of the panel, easily accessible to either occupant. The fuel shut-off was at top center, and a primer pump was furnished as standard equipment. The elevator trim control crank was located on the cabin roof.

Burnished aluminum control wheels extended through the lower part of the instrument panel, set in steel encased rubber bushings to insure smooth movement of the controls and reduce vibration. Control friction was minimized by the use of oilite bearings in the ailerons and ball bearings in the control column. Ailerons were of the Frieze type. Rudder pedals for both occupants were located on the cabin floor, and toe brakes were available on the pilots side.

The KC had a new tripod or split axle type landing gear of 70 inch tread. It was encased in a streamline fairing with oleo shock struts on each side. Wheels were the Goodyear #LMBM three-inch size, fitted with Goodyear 18x8-3 tires and tubes. Wheel brakes were available as an extra cost option. At first the plane was offered with a piston type tail skid with a cast iron shoe, but most were later changed to Aircraft Associates steerable tail wheels.

This Aeronca model was intended as an economy model, dubbed the "Scout". (Actually, few used this designation, but the "Chief" name that followed gained wide acceptance.) It was originally priced at $1590, but was later dropped to $1290 in 1939. Supposedly, the lower price of the A-40 engine made this price possible. A number of options were offered to upgrade the plane if desired by the customer. In addition to the extra door and wheel brakes, a cabin heater, floor carpeting; fire extinguisher, battery, navigation lights, tail wheel, custom upholstery, wheel pants and a twin ignition engine could be ordered at extra cost.

The Aeronca K offered more of a challenge to the student pilot than most other lightplanes. This was largely due to the higher wing loading. Whereas the Cub had 178 square feet of wing area, and the Taylorcraft 168, the Aeronca's wing had only 146 square feet. This obviously led to a higher stalling speed. Inexperienced pilots, accustomed to the forgiving characteristics of the gentle Cubs, were sometimes fatally surprised when the K spun out of a steep slow turn, often on final. This fostered a reputation, somewhat unearned, that the plane was dangerous to fly.

The Aeronca KCA which evolved into the famous "Chief "series

JUPTNER

When the KC arrived on the scene, it did nothing to improve this situation. Having about ten percent less horsepower than the E-113 equipped K, it was somewhat underpowered. On hot days, with a full load, it would hardly fly, and certainly when flown by inexperienced pilots, would be challenging. Only 35 KC's were built before the A-50 powered KCA was introduced on 3-10-38.

Thus the lack of horsepower was quickly remedied. The new Aeronca KCA, with the Continental A-50 engine, received its ATC #675 on 4-26-38, the first plane to be certificated with this new and more powerful motor. It was a stock KC in all respects, still had the small chord wing, so only firewall forward items were changed. With its up-exhaust system, the installation,

The Author's eldest son, Stan, age about 3, (now a 50+ grandfather),in the cockpit of our Aeronca K. Note the wide door, large circular tachometer on the panel and the cast aluminum control wheels.

as in all planes, was a rather awkward affair. It even retained the 10 gallon gasoline tank, hardly enough for the larger powerplant.

However, like the KC, the KCA was only a transition model, not long in production. Only 72 were built. A few months later, it was extensively modified, and became the first of the famous "Aeronca Chiefs". It had greater wing area, 168 sq ft, 12 gallons gas capacity, and a larger cabin with much more comfort and utility. Thousands of similar models were sold by Aeronca both before and after World War II.

AUTHORS NOTE

I must admit there is a soft spot in my heart for the Aeronca KC. It was the first plane I ever owned. This is the way it happened.

After I returned from service in World War II, my wife and I settled down in Sioux City Iowa, bought a home, and began raising a family. The demands of a growing family and new career made it unlikely that I could continue flying. A few times I rented a Cub or Luscombe from the local operator, the expense involved kept these flights short and infrequent. I had no expectation of ever owning my own plane.

My work at that time involved travelling through Iowa and the surrounding states, selling air conditioning and refrigeration equipment. Still interested in planes and flying, I would often detour to the country airports and visit with the operators. Those were dark days for private aviation, war surplus Stearmans and Fairchilds had flooded the market, so even new production Cubs and Taylorcrafts sold for a fraction of their original price. Still, any decent plane cost four to five hundred dollars, more than I felt I could spend on a hobby.

Then one day I was driving down the highway north of Worthington, Minnesota, and there, parked right next to the road, was a cute little monoplane tied to a fence. The ground had been ploughed around the plane so it evidently had not been flown for some time. Curious as always, I climbed the fence and looked the plane over. It appeared complete and in remarkably good condition. The registration card said it was an Aeronca KC, NC 19779, and listed the owners name and address.

Returning home that weekend, I sent a letter to the registered owner, asking if the plane was for sale; and if so, at what price. I was surprised to get a reply a few days later saying the plane was indeed for sale, asking only $150! This was a plane I could afford. I replied by return mail sending a $15 down payment and offering to pay the balance when the owner flew the plane to the nearest airport. The next week he flew the plane to Worthington, and I sent him a check for the balance on December 2, 1950. Now I owned my first plane, an Aeronca KC, NC 19779.

The first weekend of good weather we flew the plane home to Sioux City and gave it a closer inspection. It proved to be a low time airplane, recently recovered, and in excellent condition. Several years of enjoyable, and cheap, flying followed. The plane performed well, getting off easily with our small family aboard, and cruising at 80 MPH. Several times we flew to the home farm near Paullina, landing in whatever hayfield was available, taking friends and relatives for rides.

So far as I could tell, the plane had no bad habits, but it certainly had no excess of power. On takeoff you couldn't haul it into the air like a Cub; rather you had to let it run with the tail up until it decided to fly off the ground. Knowing its reputation, I was always careful to keep the nose down and airspeed up in turns. We eventually traded it for a second car, but at least we have the enjoyable memories of flying our own Aeronca KC.

Specifications
Aeronca KC
ATC #655 8-25-37

AERONCA KC AND CF, ATC 655
I - SPECIFICATIONS PERTINENT TO ALL MODELS:
Propeller — Maximum permissible diameter 81 in.
Placard speeds — Level flight or climb 95 mph True Ind.
Glide or dive 129 mph True Ind.
Fuel capacity — 10 gals. (One tank in fuselage)(-13)
Oil capacity — 1 gal. (-35)
No. pass. — 1 (+18)
Baggage — 20 lbs. (+41)
C.G. limits — (+11.4) and (+18.5)
Spec. basis — Approved Type Certificate No. 655
EQUIPMENT: (Datum is wing leading edge)(* Means net increase)

Class I:
101. 3 in. wheels (Goodyear 3LNBM, 3NBA or 3LNBA) with 16x7-3 tires
102. Tail skid
103. Propeller - wood (fixed or adj. pitch)

Class II:
200. Miscellaneous items as noted in approved weight and balance report

Class III:
301. Wheels
(a) Goodyear 3LMBM, 3LMBA or 3MBA (with brakes) — 9 lbs.* (-5)
(b) Shinn (without brakes) Net decr. 1 lb. (-5)
(c) Shinn (with brakes) — 4 lbs.* (-5)
302. 18x8-3 tires and tubes — 5 lbs.* (-5)
303. Heater (a) Cabin — 2 lbs.* (-36)
(b) Carburetor — 1 lb.* (-35)
304. Carpet — 2 lbs.* (-3)
305. Extra instruments — 10 lbs.* (-5)
306. Parachutes — 20 lbs. each
307. Auxiliary 5 gal. fuel tank (See NOTE 1) — 10 lbs.* (+25)
308. Storage battery — 13 lbs. (+44)
309. Extra door — 2 lbs. (+20)
310. Special upholstering
311. Tail wheel — (+165)
(a) Steerable (Aircraft Assoc.) with 6x2.00 tire — 4 lbs.*
(b) Swiveling (Universal Alloy H-2) with 5x1.75 tire — 3 lbs.*
(c) Swiveling (Modified Universal Alloy H-2, H-3 fork) with 6x2.00 tire — 4 lbs.*
312. Special landing gear (Dwg. 3-43 or 3-44)
313. Dual ignition engine (Continental A-40-5)(KC only) 12 lbs.*
314. Wheel streamlines — 8 lbs. (0)
315. Skis — Use actual weights
(a) Washington Aircraft 1200
(b) Federal SA-1 or SA-2
(c) Federal SC-1 or SC-2
(d) Marston MPS-1200
(e) Marston MPS-1600
(f) Jennings TY-2
(g) Richards 1-B
(h) Graf-Shupe A
(i) Heath 655

NOTE 1. Fuel arm (+45) with item 307.
NOTE 2. Eligible for export as follows, subject to inspection for equipment specified in Chapter XII of Inspection Handbook: (March 22, 1939)
A. Model KC:
(a) Canada (Continental A-40-4 engine must be placarded for 39 hp at 2500 rpm)
- Landplane
- Skiplane - not eligible. However, structure complies with Canadian requirements for ski installation when item 312 is installed. Maximum ski height 12 in.
(b) All other countries.
B. Model CF:
(a) All countries except Canada and Australia.

II - MODEL KC, 2 PCLM:
Engine — Continental A-40-4 (See also item 313)
Placard limits — Maximum, except take-off
-- in. Hg., 2575 rpm (40 hp)
Take-off (one minute)
-- in. Hg., 2575 rpm (40 hp)
Standard weight — 1060 lbs.
Serial Nos. — KC-1 and up mfrd. prior to 12-12-40 eligible. Approval expired as of that date.

Chapter Fourteen

Piper J-3

Bill Batesole's perfectly restored early 40 HP J-3 BATESOLE

By the fall of 1937, Pipers J-2 production was in full swing at the new Lock Haven factory, exceeding 100 units per month. While his business was certainly "booming", Bill Piper could see that he would need to bring out new models if he were to meet his new and growing competition. These changes were made all the more urgent by the coming introduction to the trade of several 50-65 HP aircraft engines. Continental had furnished Luscombe with a sample A-50 in July of 1937, and a new plane was being designed around it. Piper and Taylorcraft, Continental's best customers, must have received demonstrator models also.

In September of 1937, Piper gave the "go ahead" for a new model to take the 50 HP engine. To keep costs down, any changes to the basic J-2 were to be minimal. In fact, the prototype J-3 airframe was actually a J-2, S/N 1800, NX18000.

In accordance with their instructions, the engineers made a very few, hardly noticeable changes. Although their announcement advertisements trumpeted "Over 50 changes in construction and appearance", most were insignificant details. Perhaps the most visible change was the balanced rudder and the fairing of the fin into the top deck. The rear windows were enlarged and the cabin windows were made an integral part of the fuselage, not removable.

The wing plan form and aileron system remained the same, but the built-in center section was eliminated. Early in the production run, the ailerons were changed to the Frieze type. The ribs were still monel alloy; the spars were spruce and the compression members were of chrome-moly steel. The control stick mounting and arrangement was completely changed using a torque tube instead of cables and pulleys. Several parts of the fuselage structure were changed from 1020 steel to chrome-moly 4130, the obvious purpose being to allow an increase in gross weight when the new and larger motors were installed.

Announcing the New

1938 CUB SPORT

Here it is—the finest Cub ever built—a Cub that sparkles with new beauty, new comfort and new convenience! More than a score of brilliantly engineered refinements make this new 1938 addition to the Cub line the world's outstanding airplane value.

You will be delighted with the handsome new color combination available in this new 1938 Cub Sport with dual controls. And you'll like its newly appointed cabin, its deeper, wider, more comfortable seats, its additional leg room, its new stabilizer control, push-pull ignition switch, gasoline shut-off, streamlined fin and the new step that facilitates getting in and out.

The new 1938 Cub Sport cabin—whipcord upholstered—is completely weather-tight. No drafts, rain or wind can detract from your flying comfort. Its new type window raises straight up and down for greater convenience. Seats are wider and deeper. All mechanism has been neatly housed providing a clean, clear floor with additional leg room.

LOOK AT THE PRICE – ONLY $1395

Yet, with all these advanced features, the price of the new 1938 Cub Sport is extremely low — only $1395 F.A.F. Lock Haven, Pa.! And you need pay but $465 down—the balance in easy monthly payments. The price of the 1938 Cub Trainer remains unchanged—$1270 F.A.F. Lock Haven, Pa. — $425 down. Get your 1938 Cub now! Ask your dealer for a free flight demonstration in this safe airplane.

ONLY $465 DOWN

Operation of the new 1938 Cub with dual controls has been made even more simple by the provision of a stabilizer crank with indicator and of conveniently located push-pull ignition switch and gasoline shutoff. A new type of quick-action safety belt comes as standard equipment. Other new features include a rear view mirror for even greater safety and a convenient step for getting in and out of the plane.

FREE!

Send today for full particulars concerning the new 1938 Cub Sport, illustrated Cub folder and name of your Cub dealer. Use coupon at the right if you wish. Piper Aircraft Corp., 18 Aviation Street, Lock Haven, Pa.

PIPER AIRCRAFT CORPORATION
18 Aviation St., Lock Haven, Pa., U.S.A.

Please send me full particulars about the new 1938 Cub Sport, illustrated folder and name of my Cub dealer.

NAME____________________

ADDRESS____________________

CITY______________ STATE________

Aero Digest

One of the first 40 HP J-3 ads

CUB J-2 and J-3 **2-place** **ATCs 595-660**

Models J2 Cub Trainer and J-3 Cub Sport; data given in that order.
Specifications. Span 35'2.5" (10.75 m.). Length 22'5" (6.83 m.); 22'2.75" (6.77 m.). Height 6'8" (2.04 m.). Wing area 178.5 ft.^{2} (16.6 m.^{2}). Power loading 25 lbs./h.p. (11.3 kg./h.p.). Wing loading 5.6 lbs./ft.^{2} (27.4 kg./m.^{2}). Empty weight 574 lbs. (261 kg.). Useful load 426 lbs. (194 kg.). Payload 190 lbs. (86 kg.). Gross weight 1000 lbs. (453.6 kg.). Baggage 20 lbs. (9.1 kg.). Fuel 9 gals. (34.1 lit.). Oil 1 gal. (3.8 lit.).
Engine. Continental A-40-4 or A-40-5, 40 h.p. at 2575 r.p.m.; 37 h.p. at 2350 r.p.m.
Performance. Maximum speed 84 m.p.h. (135.2 k.p.h.); 86 m.p.h. (138.4 k.p.h.). Cruising speed 70 m.p.h. (112.6 k.p.h.); 72 m.p.h. (115.9 k.p.h.). Landing speed 30 m.p.h. (48.3 k.p.h.). Service ceiling 10,000 ft. (3048 m.). Climb per min. 400 ft. (121.9 m.). Cruising range 216 mi. (347.6 km.); 210 mi. (338 km.).
Construction. Fuselage: Warren truss type: welded steel tubing: fabric covered; welded steel tubing engine mount; tandem seating arrangement; dual controls; quick opening section to right of seats to permit easy entrance and exit. Wing: USA 35B airfoil section; internally braced; wood spars, fabricated aluminum alloy ribs; fabric covered. Tail group: internally braced; welded steel structure; fabric covered; stabilizer adjustable in flight. Fixed split type landing gear. Tail skid swiveling through 60°. Licensed by Department of Commerce as a seaplane on Edo model D-1070 floats.
Standard equipment. Sensenich fixed pitch propeller, Goodrich tires, Hayes wheels, Rusco endless ring shock absorbers, Pyrene fire extinguisher, Pyralin windshield and cabin enclosure, Shakespeare throttle controls. Optional equipment Freedman-Burnham variable-pitch wood propeller, battery, Grimes navigation lights, Criley-Norman compression starter.
Instruments. Aeromarine altimeter, airspeed indicator, compass; U. S. Gauge oil pressure and oil temperature gauges; Waltham tach.

Specifications AERO DIGEST

The cabin was made more user-friendly by widening the rear seat to a full 26 inches; the leg room was lengthened by 3 inches. Seats were now of welded steel tube construction, replacing the plywood frames of the J-2. With the change of seating control mechanism, the floor would be full width of varnished 1/4 inch plywood.

Instead of the primitive "sash cord" stabilizer control, an automotive style crank system was used along with a sliding position indicator. The ignition switch and the fuel shut-off control were recessed into the left hand side of the cabin wall, within easy reach of both occupants.

One quite visible and needed change was the throttle. A quadrant type replaced the sliding rod used on the E-2 and J-2. (The Author once owned a late production J-2 which had a quadrant-type throttle, so this may have been used on the last J-2's also.)

The J-3 40 HP "Cub Trainer" received its ATC #660 on 10-31-37. The first production unit was S/N 2000, NC20000. Initially, the list price was set at $1270; this was lowered to $1098 in 1939 and finally to $995 in 1940. This $995 price, ($333 down), was featured prominently in all advertising through 1940, giving Piper a low-price leader no one could match.

About 300 J-3 40 HP planes were built before they were discontinued in late 1940. The 50 HP model J-3 C-50 received its ATC # 691 on 7-14-38. This introduction of a higher-horsepower, better performing model eliminated the 40 HP planes from consideration by most customers.

The "Piper J-3 Cub" went on to become one of the most famous planes of all time. It comprised half of the CPT fleet before the start of WW II. Then, with a coat of olive drab paint, became the beloved L-4 liaison "grasshopper". When production ceased in 1947, over 12,000 had been built.

An A-40-5 installed in an early J-3. Note the boot hook and wire cowling fasteners. BATESOLE

Specifications
Piper J-3 "Sport"
ATC #660 10-31-37

PIPER J-3, 2 PCL-SM, ATC 660

Engine — Continental A-40, A-40-2, A-40-3 or A-40-4 (See also item 305 and. NOTE 4)

Placard limits — A-40, A-40-2, A-40-3) Maximum, except take-off -- in. Hg., 2550 rpm (37 hp)
Take-off (one minute) -- in.Hg., 2550 rpm (37 hp)
A-40-4) Maximum, except take-off -- in. Hg., 2575 rpm (40 hp)
Take-off (one minute) -- in. Hg., 2575 rpm (40 hp)

Propeller — Maximum permissible diameter 81 in.

Fuel capacity — 9 gals. (One tank in fuselage) (-18)

Oil capacity — 1 gal. (In engine crankcase)

No. passengers — 1 (+9) and (+36)

Baggage — Landplane 20 lbs. (+44)
Seaplane 12 lbs. (Placard compartment: "Includes anchor and rope 12 lbs. when carried.")

Weights — Empty Use actual (Seaplane approximately 88 lbs. net increase over landplane.)
Standard - Landplane 1000 lbs.
Seaplane 1070 lbs.

C.G. limits — Landplane (+10.6) and (+22.7)
Seaplane (+10.2) and (+20.8) (Edo D-1070 floats)
(+11.8) and (+23) (Edo 54-1140 floats)

Specification basis — Approved Type Certificate No. 660

Serial Nos. — 1100 thru 1200, and 1999 and up mfrd. prior to 10-15-39 eligible. Approval expired as of that date. (See NOTE 4.)

EQUIPMENT: (Datum is wing leading edge) (* Means net increase)

Class I.
(a) Landplane
101. Propeller - wooden (fixed or adjustable pitch) 9 lbs.
102. 7.00-4 wheels (Shinn) and tires 21 lbs. (+3)
103. Tail skid 4 lbs.
(b) Seaplane: Item 101 PLUS
151. Edo D-1070 float installation 145 lbs.
152. Aux. fin (on left and right stabilizer) (See NOTE 3)
153. Diagonal tension straps under front seat

Class II.
(a) Landplane
200. Miscellaneous items as noted in approved weight and balance report.
(b) Seaplane: Items 200 and 309

Class III.
(a) Landplane
301. Wheels
(a) 18x8-3 (Goodyear 3LNBM) Net decrease 2 lbs.
(b) 8.00-4 (Hayes 840) No weight change
(c) 8.00-4 (Hayes 841) with brakes 4 lbs.*
(d) Steerable tail wheel with 6 x 2.00 tire 3 lbs.*
(1) Aircraft Associates (Dwg. E-4)
(2) Aero Activities (Dwg. 1-C)
302. Battery (Reading 3-BRL-6) 8 lbs.
303. Carburetor heater 2 lbs.
304. Combination carburetor and cabin heater 3 lbs.(-39)
305. Dual ignition engine - Continental A-40-5 13 lbs.*(-39)
(Placard limits same as for A-40-4)
306. Parachutes (a) Front 20 lbs. (+9)
(b) Rear 20 lbs. (+36)

PIPER J-3 (Continued)
307. Miscellaneous extra instruments
308. Emergency exit
309. Everel single blade propeller 12 lbs.*
310. Wheel streamlines 6 lbs.
311. 12 gal. fuel tank (replacing 9 gal. standard tank)
312. Frieze type balanced ailerons No weight change
313. Skis (Use actual weight)
(a) Piper S-1000
(b) Air Transport A, 1220-480 or 1460-580
(c) Federal SA-1, SA-2, SC-1 or SC-2
(d) Marston MFS-1200 or MFS-1600
(e) Wash. Aircraft 1200
(f) Fairbanks MF-5
(g) Aviation Service B
(b) Seaplane: Items 302 to 308, inclusive, PLUS
351. Edo 54-1140 float installation (See NOTE 3) 147 lbs.

NOTE 1. Some aircraft of this model incorporate splices near the tip in the wing spars made at the factory.

NOTE 2. Eligible for export as follows, subject to inspection for equipment specified in Chapter XII of Inspection Handbook: (January 22, 1940)
(a) Canada - Landplane
Skiplane - not eligible. However, structure complies with Canadian requirements when item 313 (a) or equivalent is installed.
Seaplane - maximum standard weight 1070 lbs.
(b) Great Britain provided rear spars are reinforced in accordance with Piper Dwg. J2-L15. If 13/16 in. spars are used they must also be reinforced in accordance with Piper Dwg. J3-A-107.
(c) Australia provided that:
(1) Wing spars are reinforced in accordance with Piper Dwg. J2-L15. If 13/16 in. rear spars are used they must also be reinforced in accordance with Piper Dwg. J3-A-107.
(2) The following placard speeds are displayed:
Level flight or climb 72 mph Ind.
Glide or dive 108 mph Ind.
(d) All other countries.

NOTE 3. Item 152 unnecessary when item 351 installed.

NOTE 4. Serial No. 2309 also eligible as landplane with the following:

Engine — Franklin 4AC-150 Series 40

Placard limits — Maximum, except take-off -- in.Hg., 1875 rpm (40 hp)
Take-off (one minute) -- in.Hg., 1875 rpm (40 hp)

Propeller — Maximum permissible diameter 79 in.

Placard speeds — Level flight or climb 78 mph Ind.
Glide or dive 108 mph Ind.

NOTE: Eligible for export to all countries except Canada, Great Britain and Australia.
(Formerly known as model J3F-40. SPECIAL NOTE 11 on A.M.I.N. not pertinent.)

Chapter Fifteen

Ross RS-1

The Ross RS-1 "Parasol" JUPTNER

It was the last and certainly the least!

The little Ross RS-1 monoplane was the last plane to receive an Approved Type Certificate using the Continental A-40 engine. Just why Orrin Ross would go to the considerable effort and expense of such certification in late 1940 will probably never be known. The A-40 engine was no longer in production, the A-50 and A-65 had long ago replaced it. And twin ignition would be required, starting in the Fall of 1941.

The Ross Aircraft Company was formed at Amityville, NY by Orrin E. Ross and his wife Florence. They called their plane "The Ross Sport Parasol, Model RS-1". It was a bare minimum of airplane, as was reflected by the price, $875. Perhaps they hoped to reach a market that could not afford the $1500 Cubs and Taylorcrafts.

Certainly it had been built with low cost in mind. It was a small high-wing monoplane, built using the typical wood wing construction and steel tube fuselage. Weighing only 518 lbs., it seated two in tandem much like the earlier E-2 Cub. The landing gear was a stiff legged affair, using 16x7-4 Goodyear Airwheels for shock absorption. No brakes, just a tail skid.

Not surprisingly, the Ross plane did not sell. In fact only one was ever built, S/N 1, NX1316. After World War II the same design was offered with the Lycoming O-145-B2 engine but there is no record of any being produced.

Specifications
Ross RS-1 & RS-2
ATC-732, 8-28-40

ROSS AIRCRAFT COMPANY

MODELS: Ross RS-1 and RS-2L

T.C. NUMBER: T.C. 732

I - MODEL RS-1, 2 POLM (Approved 8/28/40)

Engine	Continental A-40-5
Fuel	73 minimum octane aviation gasoline
Engine limits	For all operations, 2575 rpm (40 hp)
Airspeed limits	Level flight or climb - 90 mph True Ind. Glide or dive - 122 mph True Ind.
C.G. range	(+20.2) to (+24.8)
Maximum weight	940 lbs.
No. seats	2 (+17, +47) (See NOTE 2.)
Maximum baggage	16 lbs. (-20)
Fuel capacity	9 gallons (-8)
Cil capacity	1 gallon (-33)
Serial numbers eligible	1 and up
Required equipment	1, 102, 201 and 202

SPECIFICATICNS PERTINENT TO ALL MODELS:

Datum	L.E. of wing
Leveling means	Level top longerons between cockpits
Certification basis	Type Certificate No. 732 (CAR 4a)
Production basis	None (See NOTE A)
Export eligibility	Eligible for export as follows, subject to provisions of MOP 2-4: (a) Canada - Landplane - eligible - Skiplane - not eligible (b) All other countries except Australia, Great Britain and New Zealand.

EQUIPMENT:
Propellers and Propeller Accessories

		RS-1
1.	Propeller - Fixed pitch wood (less hub) rated for required hp and rpm on the following limits: Static rpm at maximum permissible throttle setting - Not over 2450 rpm Not under 2275 rpm No additional tolerance permitted. Diameter - Not over 71 inches Not under 67 inches	8 lbs. (+44)
102.	Carburetor air heater	2 lbs. (-32)
103.	Engine - Lycoming O-145-B2	--

Landing Gear (See NOTE 4)

201.	Two main wheels assemblies, 7.00-4, Type III, Firestone Models 2AV or 2AVT with 4-ply rating tires.	24 lbs. (+9)
202.	Tail skid	3 lbs. (+183)
203.	Two main wheel assemblies, 16x7-3, E.L.P. Type, Goodyear Model 8LMBA with 4-ply rating 16x7-3 tires.	+3 lbs. (+9)
204.	Steerable tail wheel (Ort-Lancaster Drawings 10056 and 10057)	4 lbs. (+189)
205.	Two main wheel and brake assemblies, 6.00-6, Type III, Goodyear LSMBD (wheel assembly 511413M, brake assembly 511254A) with 6.00-6, 4-ply rating tires	--

Miscellaneous (not listed above)

601.	Cockpit canopy	--

NOTE A. Prior to original certification each aircraft must satisfactorily pass:
- (a) An inspection for workmanship, materials, and conformity before any covering, metal priming or final finish is applied. All woodwork may be varnished.
- (b) A final inspection of the completed aircraft.
- (c) A check of the flight characteristics.

Chapter Sixteen

Strangers

This Greeenwood-Yates twin, built in 1935, sports two A-40's. The engine cowls look as if they came off a J-2 Cub. This rather modern design obviously needs larger engines. JUPTNER

Because it was the only small, low-cost airplane engine available in the early 1930's, the Continental A-40 was called upon to power a number of experimental, one-of-a-kind projects. None shown in this chapter were ever certificated with the A-40, most for obvious reasons.

While there were probably dozens of such planes, the several featured on the following pages are certainly representative of these "Strangers". With the exception of the famous Ercoupe, none went into production.

The Greenwood-Yates twin engined plane, pictured above, is an example of one such experimental project. Built in Oregon in 1935, it featured an unusual geodetic all wood construction which supposedly meant that no single structural member could fail and cause a catastrophic event. It was fabric covered, with a military style tandem seating arrangement. The landing gear may have been retractable. The pair of A-40's were not powerful enough, so two Menasco M-50's were installed when they became available. This change probably did not offer much improvement in performance. Only one model, NX15546, was built.

First Ercoupe, the Model 310, flying, with a single tail and a Continental A-40. BERGEY

A later version, now with the distinctive twin tail, but still powered by the A-40 BERGEY

Most aviation historians will be surprised to learn that the first Ercoupe was powered with an A-40. But Fred Weick, who had worked out the design through 1936-37, knew a larger engine would be needed. When the plane was nearing the production stage, Weick met with Continental's Harold Morehouse who had been brought in to upgrade the A-40. Morehouse ended up designing a completely new engine, which would become the A-50, A-65 etc.

Continental's executives had still not approved the Morehouse design for production, leaving the Ercoupe people with a fine plane, but no suitable engine. Morehouse was unhappy with the Continental management because of this. In an unusual turn of events, Weick hired Morehouse to design a new engine for the Ercoupe, a 4-cylinder inverted in-line of 65 horsepower. Six of these engines were actually produced, and certification tests on the plane went forward.

A production Ercoupe, with the distinctive twin tail. This plane is being tested at the Shinn factory at Butler, PA SHINN

By the time the Ercoupe engine was ready, a number of firms had announced their entry into the 50-65 HP engine field. Continental, Lycoming , Franklin and Menasco had all introduced their higher horsepower engines, so the costly Ercoupe-produced model was no longer economical to use. Actual certification of the Ercoupe was completed with a Continental A-65 in 1940.

This factory photo of a Skycycle in flight was taken in March of 1945, after the Lycoming O-145 engine was substituted for the Continental A-40 — PIPER MUSEUM

Not all the unusual applications of the Continental A-40 were made by wild-eyed shadetree visionaries. Even the respected Piper organization ventured into uncharted waters with their little "Skycycle".

As WWII wound down in the later part of 1944, most light aircraft manufacturers were searching for possible postwar designs. One of Piper's experiments was a very simple , one-place plane, whose fuselage was made from a Corsair fighter drop tank. Piper had been making these tanks by the thousands during the war, so they were readily available and very cheap. As first designed, the plane was a mid-wing monoplane, but this configuration was almost immediately changed to a low wing. Wanting the lightest powerplant possible, an A-40 engine was pulled out of storage and installed behind an open cowl in the nose.

The metal tail cone was fastened to the drop tank which served as the cockpit; a short, tapered wing plan was used, fabric covered. A wide-track conventional landing gear was fitted. The plane was very small, 15' 8" long; had a wing span of 20', and a gross weight of only 630 pounds.

From the start the Skycycle exhibited some decidedly unpleasant handling characteristics, which were never entirely eliminated by the addition of large wing fillets and other modifications. After using the Continental A-40 for the first tests, a Lycoming O-145 engine was installed. Perhaps the greatest drawback was the single seat arrangement; no real market could be envisioned for such a plane. Bill Piper Sr. cancelled the project in early 1946.

Everly "Whiffle Hen" JUPTNER

The Everly "Whiffle Hen", shown above, was built by the Everly School of Aeronautics at Salem OR.It was powered by one of the very early production models of the A-40. Seating two people, side-by-side, it is similar in appearance to the Aeronca C-3 and the later produced Taylorcraft A.

It had a pleasing appearance, the performance should have been adequate, but since it was built only as school project, to give students practical experience, no volume production was likely considered. The "Whiffle Hen" name certainly did not add much to its image.

Little is known about this Stirling-Edwards twin, shown in this 1938 photo.. It was reported to be first powered by two A-40's. These were later replaced by two A-50's. JUPTNER

The Mattley "Flivver Plane" JUPTNER

The Mattley monoplane, shown above evidently sported one of the very first A-40's produced by Continental. This photo is dated 2-22-31, which even pre-dates the motor's ATC approval. It was common, however, for manufacturers to send sample motors to prospective customers before such approval was obtained.

This rather crude high-wing monoplane was built as a school project, to give students practical experience. It was first offered, as the FP-1, NX10615, (a single place model) in 1930-31. The second model, slightly enlarged to seat two, became the FP-2, NX12707. The plane was seen flying, at least until 1935, in the San Bruno, CA, area.

This Zueck "Planemobile" appears to be an attempt to build a roadable plane. The struts allow for folding wings, and the third wheel is steerable. JUPTNER

The "Arup" flying wing, NX12894

MALOY

This odd-looking contraption was reported to have been test flown by Jimmy Doolittle's nephew in South Bend, Indiana. It had an unusual, all-wing design with a thick, low aspect ratio planform. The pilot was in an almost horizontal position; only his head showing above the airfoil.

At the center of the trailing edge, small flaps worked as elevators, while further out on the wing, another small set acted as ailerons. The rudder and fin appear conventional. The plane could land at a steep, nose-high attitude, rolling only about 60 feet. Only one of this model was produced.

Chapter Seventeen

The Legacy

The A-40's successor, the Continental A-50. It had an "up" exhaust arrangement, required a separate oil tank, and sported a starter and generator. CONTINENTAL

By late 1936, it was evident to Continental's management that the days of the trusty A-40 were nearly over. Since 1931, when it was first introduced, the little engine had been the mainstay of the light aircraft industry. A dozen planes, representing a major portion of our country's light aircraft fleet, had been certified for its installation. A number of other planes used it for experimental purposes. Except for the two-cylinder Aeronca engine, the A-40 had no real competition in the low horsepower field.

But the situation in this market was changing. Lightplanes were becoming more sophisticated, with features added to give more comfort, safety and speed. These additions, such as brakes, additional instruments, and more comfortable cabins, naturally added to the empty weight. Customers were demanding faster, more practical planes with additional load carrying capacity, but the little A-40, limited to 40 HP maximum, could do no more.

Adding to the pressure for change were the rumors that other manufacturers were about to enter this field. The defunct Franklin Auto Co, of Syracuse, NY, had been purchased by a group of former employees. They were designing a 4-cylinder flat aircraft engine to directly challenge Continental in this field. Lycoming, who had been making engines for several auto manufacturers, was also looking at this market. Even Al Menasco, making his fine in-line's on the west coast, was designing a small flat four. To hold on to this market, Continental clearly needed a higher horsepower engine.

It appears that the very small size of this market, compared to the numbers of engines used for industrial applications, kept Continental from making a decision earlier. Even though the A-40 was the industry's most popular engine, only about 3000 had been sold by the spring of 1937. Their larger aircraft engine, the R-670 air-cooled radial, sold fewer than 100 units per year.

They had, however, been interested enough to assign the task of designing a new, larger, engine to their resident expert, Harold A. Morehouse. Morehouse had been hired in 1933, on the recommendation of Robert Insley. Early in the redesign process, Morehouse determined that the L-head layout would have to be discarded, in favor of the valve-in-head arrangement. Actually, for over 10 years, Morehouse had been designing air cooled, valve-in-head small aircraft engines. (See pages 3-4 in Chapter 1)

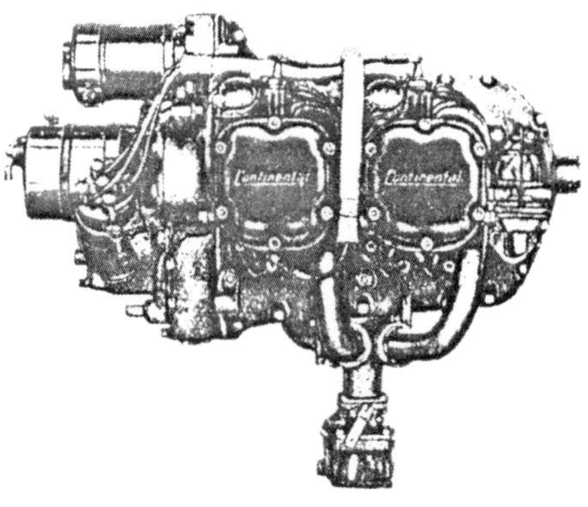

CONTINENTAL A-50 **4-cylinder — opposed** **ATC 190**

Specifications. Direct drive, air-cooled. Department of Commerce and manufacturer's rating 50 h.p. at 1900 r.p.m. Compression ratio 5.4:1. Bore 3.875"; stroke 3.625"; displacement 171 cu. in. Length 30.41"; height 23"; width 31.69". Weight (with carburetor and magneto) 150 lbs. Weight in lbs./h.p. at sea level 3. Fuel consumption at rated h.p. .55 lb./b.h.p./hr. Oil consumption at rated h.p. .015 lb./b.h.p./hr. Fuel octane rating 65 (min.).
Construction. Crankcase aluminum alloy cast; two-piece; fastened in vertical lengthwise plane through crankshaft. Crankshaft one-piece forging with 3 main bearings; four-throw type. Pistons Lo-Ex alloy, trunk type; full floating piston pin; 2 compression and 2 oil rings to each cylinder. Cast aluminum alloy heads screwed and shrunk to forged steel barrels. Connecting rods forged steel with split big end. Steel backed, cadmium-lined replaceable bearings. Fuel pump drive located front right hand of crankcase; other drive at rear. Lubrication forced feed to crankshaft main and connecting rod bearings, tappets, rocker arms and rocker housings; gear scavenge or gravity drain. Steel intake and Austenitic steel exhaust valves.
Accessories. Scintilla magneto, Bendix-Stromberg carburetor, B.G. spark plugs, AC fuel pump, American Hammered piston rings. Optional equipment includes Eclipse or Auto-Lite starter and generator.

A-50 specifications as published in 1938 Aero Digest

The resulting design, first shown publicly in late 1937, was a larger flat four, the Continental A-50. It was a substantial departure from the A-40 arrangement. The crankcase was a two-piece heat-treated aluminum alloy casting, bolted together at the vertical lengthwise plane. Rigid transverse webs held the three crankshaft main bearings and the camshaft journals. Guides for the hydraulic self-adjusting tappets were cast and machined below and parallel to the cylinders. A cast aluminum cover plate was bolted onto the rear of the engine, enclosing the gear train for the magnetos, oil pump and camshaft. The connecting rod and main bearings were steel backed babbitt replaceable bearing shells.

The cylinders were of machined steel with cast aluminum heads screwed and shrunk on. Heat dissipating fins were closely spaced on both. Bronze spark plug inserts were screwed in, while the bronze intake seats and steel exhaust seats were shrunk in place. The rocker boxes were cast with the heads and sealed with stamped steel covers. The push rods were enclosed and carried lubricating oil to the valve mechanism.

Oil was pumped to all bearings through the drilled crankshaft. The first models were offered without an oil sump, but soon a rounded sheet steel sump was attached directly to the bottom of the crankcase.

When first introduced, after receiving ATC #190, the A-50 was rated at 50 HP turning 1900 RPM. In actual practice it delivered somewhat more and was received with great enthusiasm by the various light aircraft manufacturers. They knew the A-40 was not powerful enough to allow good performance; indeed some, including Luscombe, could not use the A-40 at all.

The first ad for the Luscombe "50" which appeared in March of 1938. This plane was the first plane to fly with the A-50 engine on December 17, 1937 Aero Digest

One of the first BC-50 Taylorcrafts. A beautiful plane, but the "up" exhaust is ugly. ALLIANCE LIBRARY

However, the A-50 had one glaring fault, the exhaust was vented on the upper side of the cylinder. All manufacturers were now using down exhaust systems, to keep the gasses away from the cockpit, so it is a mystery why Morehouse would use this arrangement. True, it probably scavenged slightly better, but what a price to pay. The first models of the Cub, Aeronca, Taylorcraft and Luscombe all show ugly up-exhaust arrangements.

It appears that Luscombe was the first manufacturer to actually install and fly a Continental A-50. On December 17, 1937, their prototype Model 50, (later the Model 8), was flown at the West Trenton, New Jersey, plant. The performance exceeded expectations; finally an honest 100 MPH cruise could be obtained in a lightplane. As often happened, certification was a slow process. Aeronca came first with their KCA, ATC #675 on 4-26-38. The Cub, Luscombe and Taylorcraft followed soon after.

All of these early aircraft models exhibit the ugly up exhaust arrangements; it is not surprising that a change was made to the "down" exhaust within a few months. However, the basic design proved sound and easily upgraded. After a short time the A-65 was offered, giving 65 HP at 2300 RPM. It was this Continental A-65 that became the most popular light aircraft motor of those times; powering many models of lightplanes, including most of the liaison planes, (L-2, L-3 and L-4), used by the United States Army in World War II. This basic engine went through a number of improvements, finally topping out as the O-200 rated at 100 HP.

"40" OR "50"

TAKE YOUR CHOICE!

For more than six years Cub airplanes have been powered by the dependable Continental A-40 engine. However, with the introduction and approval of the Cub with 50 horsepower, the high performance of this popular ship is greatly enhanced—quicker take-off, faster climb, more speed. Now, in addition to the regular 40 horsepower models, the Cub is available with the following 50 horsepower engines:

CONTINENTAL "50"

FRANKLIN "50" **LYCOMING "50"**

LENAPE "50" **MENASCO "50"**

In 1938, the Cub was offered with five choices of 50 HP engines! AERO DIGEST

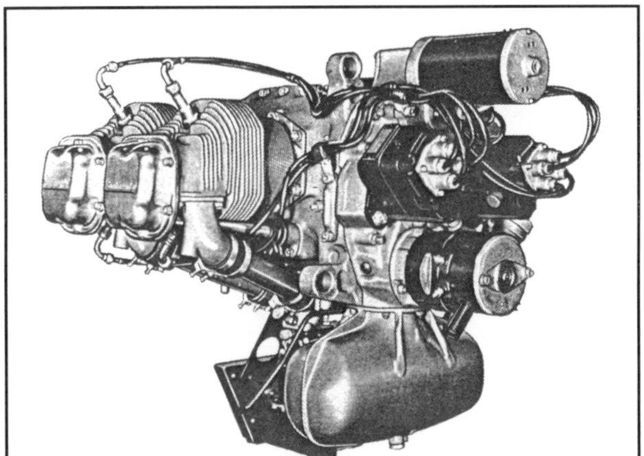

The Continental O-200 CONTINENTAL

LYCOMING 0-145 **4-cylinder** **ATC Pending**

Specifications. Direct drive, air-cooled. Manufacturer's rating 50 h.p. at 2300 r.p.m. Compression ratio 5.75:1. Bore 3.62"; stroke 3.5"; displacement 144.5 cu. in. Length 23.75"; height 19.78"; width 29.5". Weight (with propeller hub rear flange, magneto, ignition assembly, spark plugs, hub attaching parts, carburetor) 150 lbs. Weight per h.p. 3 lbs. Fuel consumption at cruising h.p. .60 lb./-b.h.p./hr. Oil consumption .025 lb./b.h.p./hr. Fuel octane rating 73.

Construction. Crankcase two-piece unit with integral cylinder blocks, split vertically through centerline and joined by studs, nuts and capscrews. Crankshaft one-piece forged alloy steel, four throw type, propeller hub rear flange forged integral. Cylinder assemblies cast aluminum heads attached to barrels by studs and nuts; cylinders integral with crank-case sections; cooling fins cast directly on heads and barrels. Pistons Lo-Ex aluminum alloy, ribbed on underside of head and machined for one oil and two compression rings. Connecting rods forged steel split cap type with bronze busings in piston end and babbitt-lined bearings in crankpin end. Replaceable bearings, three precision type main, steel back babbitt-lined heavy wall. Tachometer drive at rear of engine. Lubrication by pressure oil pump, wet sump; automatic valve gear lubrication.

Accessories. Scintilla magneto, Bendix-Stromberg carburetor, B.G. or Champion spark plugs, Thompson poppet valves.

FRANKLIN Aircooled AC-150 **4-cylinder** **ATC Pending**

Specifications. Direct drive, air-cooled. Manufacturer's rating 50 h.p. at 2400 r.p.m. Compression ratio 6.5:1. Bore 3.625"; stroke 3.625"; displacement 150 cu. in. Length 27.5"; height 18.37"; width 34.37". Weight (dry) 154 lbs. Weight in lbs. /h.p. 3.

Construction. Crankcase aluminum alloy; removable top cover; houses three main bearings. Crankshaft forged alloy steel; one-piece; four-throw; drilled for lightness and lubrication of pins and main bearings; propeller inner flange forged integral. Pistons aluminum alloy cast with full floating pins restrained by alloy steel spring at each end. Cylinder individually cast with valve rocker box integral; Y alloy. Connecting rods alloy steel with bronze bushing pressed in upper end; big end bearing replaceable thin steel filled type with copper lead lining. Celeron timing gears. Valve seats and exhaust valve guides Ni-resist cast iron. Cooling fins on cylinder head carried to exhaust port. Rocker arms operated through bronze bushings on hollow hardened pins. Main bearings interchangeable type, steel backed with babbitt lining. Lubrication full pressure throughout including valve rockers; wet sump. Automatic valve adjustment obtained by use of Wilcox-Rich hydraulic valve lifters. Oil pump incorporates aluminum housings with duralumin gears.

Accessories. Scintilla magnetos, Stromberg carburetor, Wilcox-Rich valves, Perfect Circle and Simplex piston rings.

MENASCO M-50 **4-cylinder-horizontally opposed** **ATC 191**

Specifications. Direct drive, air-cooled. Manufacturer's rating 50 h.p. at 2500 r.p.m. Compression ratio 5.25:1. Bore 3.5"; stroke 3.75"; displacement 144.4 cu. in. Length 28.25"; height 21.156"; width 28.37". Weight (with carburetor and magneto) 164 lbs. Weight in lbs. /h.p. at sea level 3.3. Fuel consumption at rated h.p. .65 lb./b.h.p./hr. Oil comsumption at rated h.p. .025 lb./b.h.p./hr. (max.). Fuel octane rating 73.

Construction. Crankcase one-piece cast aluminum alloy with cooling fins. Crankshaft steel forging, one-piece; three-throw type; supported on two main bronze bearings. Pistons cast aluminum alloy, full skirt type; full floating piston pin; each piston equipped with three compression and one oil ring above pin. Cylinder assemblies consist of nickel cast iron cylinders cast in pairs and cast aluminum alloy heads with large cooling fin area. Connecting rods forged steel, with bronze wrist pin bearings and split steel backed bearing in big end. Lubrication wet sump with gear type pressure pump supplying oil to main bearings, connecting rod bearings and camshaft bearings. Carburetion altitude control type, updraught carburetor. Valves hardened chrome nickel steel intake, heat resistant steel exhaust. Enclosed valve mechanism.

Accessories. Bendix-Stromberg carburetor, Scintilla magneto, B.G. or Champion spark plugs, Jadson valves.

AERO DIGEST

Continental faced competition in its race for more horsepower in 1938. Lycoming, Franklin, and Menasco all announced their 50 HP models

The Legacy

The Continental A-40 leaves a rich legacy. With very few exceptions, every piston engine installed in a new plane today anywhere in the world is of the opposed type. It has displaced all other design arrangements simply because it offered engineers the optimal package of power, weight, streamlining and reliability.

When reaching its final development stage in the 1980's, the opposed engine was delivering over 400 horsepower, effectively eliminating all radials from applications of this size. With the development of smaller turbine engines, 600 HP and above, the radial engine lost this market also.

The opposed motor is much smoother than a similar sized radial for the reason that its opposite pistons come together and move outward at the same time. In the later stages of development, it was able to reach 2700-2800 RPM giving a very good power-weight ratio. True, propeller efficiencies suffered, but this was somewhat offset by the introduction of controllable propellers in the small horsepower range.

Streamlining and the aerodynamic efficiencies of fuselage shapes, also favored the flat engine. With most planes using side-by-side seating, the engine cowling naturally flowed into the cabin shape. Visibility was improved in almost all cases, especially when the tricycle gear was adopted by most manufacturers. No one wishes a return to the "blind" situation in which a pilot finds himself when landing a tailwheel plane behind a large radial.

All the developments in aircraft engines mentioned above would not have been possible without the little Continental A-40 leading the way. Its introduction in 1931 represented a historic milestone in American aviation. For more than six years it remained alone on the scene, the only really practical powerplant available to the low-horsepower market. Together with the Cubs and Taylorcrafts it powered, the A-40 lifted those eager young pilots of the 1930's "into the air". It deserves a special place in our aviation history.